Australian Breastfeeding + Lactation Research and Science Translation Conference 2023

Australian Breastfeeding + Lactation Research and Science Translation Conference 2023

Editors

Zoya Gridneva
Donna T. Geddes
Debra J. Palmer
Nicolas L. Taylor
Jacki L. McEachran

MDPI

Basel • Beijing • Wuhan • Barcelona • Belgrade • Novi Sad • Cluj • Manchester

Editors
Zoya Gridneva
The University of Western Australia
Perth
Australia

Donna T. Geddes
The University of Western Australia
Perth
Australia

Debra J. Palmer
Telethon Kids Institute
Perth
Australia

Nicolas L. Taylor
The University of Western Australia
Perth
Australia

Jacki L. McEachran
The University of Western Australia
Perth
Australia

Editorial Office
MDPI
St. Alban-Anlage 66
4052 Basel, Switzerland

This is a reprint of articles from the Proceedings published online in the open access journal *Proceedings* (ISSN 2504-3900) (available at: https://www.mdpi.com/2504-3900/93/1).

For citation purposes, cite each article independently as indicated on the article page online and as indicated below:

Lastname, A.A.; Lastname, B.B. Article Title. *Journal Name* **Year**, *Volume Number*, Page Range.

ISBN 978-3-7258-0787-1 (Hbk)
ISBN 978-3-7258-0788-8 (PDF)
doi.org/10.3390/books978-3-7258-0788-8

Cover image courtesy of The University of Western Australia

Contents

proceedings MDPI

Editorial

Statement of Peer Review

Jacki L. McEachran [1,2,3,*], Zoya Gridneva [1,2,3], Debra J. Palmer [4,5], Nicolas L. Taylor [1], Ching Tat Lai [1,2,3], Lisa F. Stinson [1,2,3], Sharon L. Perrella [1,2,3] and Donna T. Geddes [1,2,3]

1 School of Molecular Sciences, The University of Western Australia, Crawley, WA 6009, Australia; zoya.gridneva@uwa.edu.au (Z.G.); nicolas.taylor@uwa.edu.au (N.L.T.); ching-tat.lai@uwa.edu.au (C.T.L.); lisa.stinson@uwa.edu.au (L.F.S.); sharon.perrella@uwa.edu.au (S.L.P.); donna.geddes@uwa.edu.au (D.T.G.)
2 ABREAST Network, Perth, WA 6000, Australia
3 UWA Centre for Human Lactation Research and Translation, Crawley, WA 6009, Australia
4 Telethon Kids Institute, The University of Western Australia, Nedlands, WA 6009, Australia
5 School of Medicine, The University of Western Australia, Crawley, WA 6009, Australia
* Correspondence: jacki.mceachran@uwa.edu.au

When submitting conference proceedings to the journal *Proceedings*, the volume's editors notify the publisher that they carried out a peer review of all published papers. Reviews are conducted by expert referees while upholding all the professional and scientific standards expected of the *Proceedings* journal.

- Type of peer review: single-blind.
- Conference submission management system: email.
- Number of submissions sent for review: 30.
- Number of submissions accepted: 30.
- Acceptance rate (number of submissions accepted/number of submissions received): 100%.
- Average number of reviews per paper: 2.
- Total number of reviewers involved: 4.
- Any additional information on the review process: not applicable.

This publication collates the proceedings of the ABREAST Conference held on 10 November 2023 in Perth, Australia.

The ABREAST Conference showcases the significant steps taken in advancing human lactation research. The collaboration between the ABREAST Network and the Geddes Hartmann Human Lactation Research Group, as well as the launch of The University of Western Australia's Centre for Human Lactation Research and Translation, reflect a commitment to promoting knowledge, innovation, and positive impacts in the field.

The conference featured keynote speakers from Australia, the United States of America, and the United Kingdom, providing a global perspective and contributing to a diverse and comprehensive exploration of topics related to human lactation. The range of research topics covered, from breastfeeding after Caesarean birth to milk composition and production, colostrum, human milk, allergy prevention, and the impact of milk on infant growth and development, illustrated a holistic approach to understanding and supporting breastfeeding.

The ABREAST Network emphasises inclusivity, diversity, passion, and innovation in lactation research, demonstrating a commitment to making a positive difference in the lives of mothers, infants, and communities worldwide. The interactive, hybrid platform provided by the ABREAST Conference has successfully created an environment for researchers, students, clinicians, and the community to exchange knowledge and insights.

The conference was well received, with a total of 161 registrations; 30 manuscripts, including invited and keynote speakers, were considered and accepted for presentation, and 23 manuscripts were published. This publication will serve as a valuable resource for those interested in advancing their understanding of human lactation.

Citation: McEachran, J.L.; Gridneva, Z.; Palmer, D.J.; Taylor, N.L.; Lai, C.T.; Stinson, L.F.; Perrella, S.L.; Geddes, D.T. Statement of Peer Review. *Proceedings* **2023**, *93*, 23. https://doi.org/10.3390/proceedings2023093023

Published: 28 February 2024

Overall, the ABREAST Conference was a successful and impactful event, fostering collaboration, knowledge exchange, and the advancement of research in the field of human lactation.

Organising Committee

Ms. Jacki L. McEachran, The University of Western Australia, Australia.
Dr. Zoya Gridneva, The University of Western Australia, Australia.
Prof. Donna T. Geddes, The University of Western Australia, Australia.
Dr. Lisa F. Stinson, The University of Western Australia, Australia.
Dr. Sharon L. Perrella, The University of Western Australia, Australia.
Dr. Ching Tat Lai, The University of Western Australia, Australia.

Scientific Committee

Dr. Zoya Gridneva, The University of Western Australia, Australia.
Assoc/Prof. Debra J. Palmer, Telethon Kids Institute, Australia.
Dr. Nicolas L. Taylor, The University of Western Australia, Australia.
Prof. Donna T. Geddes, The University of Western Australia, Australia.

Chairs

Session I

Prof. John P. Newnham, AM, The University of Western Australia, Australia.
Prof. Donna T. Geddes, The University of Western Australia, Australia.
Dr. Lisa F. Stinson, The University of Western Australia, Australia.

Session II

Dr. Sharon L. Perrella, The University of Western Australia, Australia.
Clinical Assoc/Prof. Mary Sharp, King Edward Memorial Hospital, Australia.

Session III

Dr. Lisa F. Stinson, The University of Western Australia, Australia.
Assoc/Prof. Debra J. Palmer, Telethon Kids Institute, Australia.

Session IV

Prof. Donna T. Geddes, The University of Western Australia, Australia.
Dr. Zoya Gridneva, The University of Western Australia, Australia.

Technical Chairs

Dr. Ching Tat Lai, The University of Western Australia, Australia.
Ms. Jacki L. McEachran, The University of Western Australia, Australia.

Registration Chairs

Ms. Ashleigh H. Warden, The University of Western Australia, Australia.
Ms. Grace C. McLoughlin, The University of Western Australia, Australia.
Ms. Jacki L. McEachran, The University of Western Australia, Australia.

Conference Judges—Student Poster Presentations

Dr. Alexandra D. George, The Baker Institute, Australia.
Assoc/Prof. Debra J. Palmer, Telethon Kids Institute, Australia.
Dr. Stuart A. Prosser, Western Obstetrics, Australia.
Dr. Luoyang Ding, The University of Western Australia, Australia.
Prof. Valerie Verhasselt, The University of Western Australia, Australia.
Clinical Assoc/Prof. Mary Sharp, King Edward Memorial Hospital, Australia.
Ms. Holly Tickner, Next Bite Feeding, Australia.

Student Prize Winners—sponsored by *Nutrients* (MDPI)

First Prize—Maheshwar Bhasin, The University of Western Australia, Australia.
Second Prize—Sophie A. Hughes, The University of Western Australia, Australia.
Third Prize—Jie Ma, The University of Western Australia, Australia.

Editors

Dr. Zoya Gridneva, The University of Western Australia, Australia.
Prof. Donna T. Geddes, The University of Western Australia, Australia.
Assoc/Prof. Debra J. Palmer, Telethon Kids Institute, Australia.
Dr. Nicolas L. Taylor, The University of Western Australia, Australia.
Ms. Jacki L. McEachran, The University of Western Australia, Australia.

Proceedings Compilation Team

Dr. Zoya Gridneva, The University of Western Australia, Australia.
Ms. Jacki L. McEachran, The University of Western Australia, Australia.

Conflicts of Interest: The authors declare no conflict of interest.

proceedings

MDPI

Abstract

Immediate Point-of-Care Testing of Breastmilk Sodium and Potassium Concentrations in Women with Mastitis †

Emma L. Heron [1,2,*], Ching Tat Lai [1,3,4], Leanda J. McKenna [2], Adelle M. McArdle [5] and Donna T. Geddes [1,3,4]

1 School of Molecular Sciences, The University of Western Australia, Crawley, WA 6009, Australia; ching-tat.lai@uwa.edu.au (C.T.L.); donna.geddes@uwa.edu.au (D.T.G.)
2 School of Allied Health, Curtin University, Bentley, WA 6102, Australia; l.mckenna@curtin.edu.au
3 ABREAST Network, Perth, WA 6000, Australia
4 UWA Centre for Human Lactation Research and Translation, Crawley, WA 6009, Australia
5 Monash Rural Health, Monash University, Churchill, VIC 3842, Australia; adelle.mcardle@monash.edu
* Correspondence: emma.duff@postgrad.curtin.edu.au

† Presented at the Australian Breastfeeding + Lactation Research and Science Translation Conference (ABREAST Conference 2023), Perth, Australia, 10 November 2023.

Abstract: Sodium (Na) and potassium (K) concentrations in breastmilk are often used as biomarkers to define mastitis in lactating women and can be measured with small portable point-of-care ion-selective electrodes (ISEs). The aim of this study was to test the ISEs at the point of care for accuracy and acceptability in women with mastitis. Up to 5 mL of expressed breastmilk from the affected breast of 43 women with mastitis was collected at three timepoints (day 1, 3, and 10). Immediate Na and K ISE testing was later compared to the laboratory measure of inductively coupled plasma–op-tical emission spectrometry (ICP-OES). The results revealed a statistically significant difference in Na and K concentrations between the point-of-care and laboratory testing (both $p = 0.001$, Wilcoxon signed-rank test); however, the difference was not statistically significant when compared for Na:K ratio ($p = 0.49$, Wilcoxon signed-rank test). The Bland–Altman limits of agreement were acceptable, with the majority of measurements lying within two standard deviations of the mean (Na: 94%; K: 95%; and Na:K: 96%). The testing techniques were significantly correlated for Na ($R^2 = 0.79$, $p = 0.001$) and Na:K ($R^2 = 0.99$, $p = 0.001$). Overall, participants rated the ISE point-of-care testing as very acceptable. In conclusion, immediate ISE point-of-care testing for breastmilk Na:K ratio in women with mastitis is clinically accurate and acceptable.

Keywords: breastmilk; biomarkers; sodium; potassium; ion-selective electrode; point-of-care; inductively coupled plasma–optical emission spectrometry; lactation; mastitis; breast inflammation

Citation: Heron, E.L.; Lai, C.T.; McKenna, L.J.; McArdle, A.M.; Geddes, D.T. Immediate Point-of-Care Testing of Breastmilk Sodium and Potassium Concentrations in Women with Mastitis. *Proceedings* **2023**, *93*, 1. https://doi.org/10.3390/proceedings2023093001

Academic Editors: Debra J. Palmer and Nicolas L. Taylor

Published: 19 December 2023

Author Contributions: Conceptualization, E.L.H., L.J.M., A.M.M. and D.T.G.; methodology, E.L.H., C.T.L., L.J.M. and D.T.G.; validation, E.L.H. and C.T.L.; formal analysis, E.L.H. and C.T.L.; investigation, E.L.H.; resources, C.T.L. and D.T.G.; data curation, E.L.H. and C.T.L.; writing—original draft preparation, E.L.H. and C.T.L.; writing—review and editing, E.L.H., C.T.L., L.J.M., A.M.M. and D.T.G.; supervision, C.T.L., L.J.M., A.M.M. and D.T.G.; funding acquisition, E.L.H. All authors have read and agreed to the published version of the manuscript.

Funding: This research was funded by Mercy Health Small Grants 2021/2022 and 2022/2023. Author and Ph.D. candidate, E.H., received an Australian Government Research Training Program Scholarship (https://www.dese.gov.au/research-block-grants/research-training-program). The funders had no role in the design of the study; in the collection, analyses, or interpretation of data; in the writing of the manuscript; or in the decision to publish the results.

Institutional Review Board Statement: The study was conducted in accordance with the Declaration of Helsinki and approved by the Mercy Health Human Research Ethics Committee (2021-007, 6 May 2021).

Informed Consent Statement: Informed consent was obtained from all subjects involved in the study.

Data Availability Statement: Data sharing not applicable.

Acknowledgments: Medela AG (Switzerland) is acknowledged for their kind loan of a hospital-grade breast pump; Angela Jacques for her statistical analysis assistance; and Xiaojie Zhou and Oscar Del Borrello for their technical support.

Conflicts of Interest: D.T.G. declares participation in the Scientific Advisory Board of Medela AG. C.T.L. and D.TG. receive a salary from an unrestricted research grant from Medela AG, administered by The University of Western Australia. The funders had no role in the design of the study; in the collection, analyses, or interpretation of data; in the writing of the manuscript; or in the decision to publish the results. All other authors declare no conflict of interest.

proceedings

Abstract

Human Milk Microbiome Is Altered in Mothers with Gestational Diabetes Mellitus †

Sophie A. Hughes [1,2,3,4], Sharon L. Perrella [1,2,3], Demelza J. Ireland [4], Donna T. Geddes [1,2,3] and Lisa F. Stinson [1,2,3,*]

1 School of Molecular Sciences, The University of Western Australia, Crawley, WA 6009, Australia; 22876633@student.uwa.edu.au (S.A.H.); sharon.perrella@uwa.edu.au (S.L.P.); donna.geddes@uwa.edu.au (D.T.G.)
2 ABREAST Network, Perth, WA 6000, Australia
3 UWA Centre for Human Lactation Research and Translation, Crawley, WA 6009, Australia
4 School of Biomedical Sciences, The University of Western Australia, Crawley, WA 6009, Australia; demelza.ireland@uwa.edu
* Correspondence: lisa.stinson@uwa.edu.au
† Presented at the Australian Breastfeeding + Lactation Research and Science Translation Conference (ABREAST Conference 2023), Perth, Australia, 10 November 2023.

Keywords: gestational diabetes mellitus; human milk microbiome; body mass index; milk production

Gestational diabetes mellitus (GDM) is a metabolic disease of pregnancy that is associated with alterations in the maternal and infant gut microbiota. The intake of human milk shapes the infant gut microbiome; however, to date, only one small study has investigated the impact of GDM on the milk microbiome. Therefore, we aimed to add more robust data to this field by characterising the human milk microbiome of mothers with GDM and without GDM over the first 6 weeks postpartum. Given the relationships between body mass index (BMI) and milk production with GDM, our secondary aims examined relationships between maternal BMI, milk production and the milk microbiome. Eighty-three mothers were included in the analysis (forty-three with GDM and forty without GDM). Participants measured their milk production at 3 weeks postpartum by test weighing their infants before and after each feed over 24 h and supplied milk samples at 1, 3 and 6 weeks postpartum. Full-length 16S rRNA gene sequencing was performed. Milk from mothers with GDM had a higher alpha diversity than milk from mothers without GDM (richness, $p = 0.026$; Shannon diversity, $p = 0.044$). Beta diversity differed between the two groups (PERMANOVA, $p = 0.034$). At the OTU level, the composition of the milk microbiome varied significantly based on GDM status and maternal pre-pregnancy BMI. Mothers with GDM were more likely to have low milk production (<600 g/24 h; $p = 0.018$). However, the milk microbiome was not associated with milk production. This study contributes further to our understanding of differential microbiome outcomes in relation to GDM.

Citation: Hughes, S.A.; Perrella, S.L.; Ireland, D.J.; Geddes, D.T.; Stinson, L.F. Human Milk Microbiome Is Altered in Mothers with Gestational Diabetes Mellitus. *Proceedings* **2023**, *93*, 2. https://doi.org/10.3390/proceedings2023093002

Published: 19 December 2023

Author Contributions: Conceptualization, L.F.S., S.L.P. and D.T.G.; methodology, L.F.S.; formal analysis, S.A.H.; investigation, S.A.H.; resources, D.T.G.; data curation, S.L.P. and S.A.H.; writing—original draft preparation, S.A.H.; writing—review and editing, L.F.S. and D.T.G.; visualization, S.A.H.; supervision, L.F.S., D.T.G. and D.J.I.; funding acquisition, D.T.G. All authors have read and agreed to the published version of the manuscript.

Funding: This research was funded by an unrestricted research grant from Medela AG (Switzerland).

Institutional Review Board Statement: The study was conducted in accordance with the Declaration of Helsinki, and was approved by the Human Research Ethics Committee of The University of Western Australia (RA/4/20/5657).

Informed Consent Statement: Informed consent was obtained from all subjects involved in the study.

Data Availability Statement: Sequence data have been submitted to the NCBI SRA (BioProject Submission: SUB13895764).

Acknowledgments: The authors would like to acknowledge Ashleigh Warden for collecting the samples and Matthew Payne for the use of his laboratory space and equipment.

Conflicts of Interest: D.T.G. declares participation in the Scientific Advisory Board of Medela AG. D.T.G., S.L.P. and L.F.S. are/were supported by an unrestricted research grant from Medela AG, administered by The University of Western Australia. The funders had no role in the design of the study; in the collection, analyses, or interpretation of data; in the writing of the manuscript; or in the decision to publish the results.

proceedings

MDPI

Abstract

The Development of the Human Milk Microbiota over the First Two Years Postpartum in the Breastfeeding Longitudinal Observational Study of Mothers and Kids (BLOSOM) Cohort †

Ruomei Xu [1,2,3], Ali S. Cheema [4], Mark P. Nicol [5], Donna T. Geddes [1,2,3] and Lisa F. Stinson [1,2,3,*]

[1] School of Molecular Sciences, The University of Western Australia, Crawley, WA 6009, Australia; ruomei.xu@research.uwa.edu.au (R.X.); donna.geddes@uwa.edu.au (D.T.G.)
[2] ABREAST Network, Perth, WA 6000, Australia
[3] UWA Centre for Human Lactation Research and Translation, Crawley, WA 6009, Australia
[4] Telethon Kids Institute, Nedlands, WA 6009, Australia; alisadiq.cheema@telethonkids.org.au
[5] School of Biomedical Sciences, The University of Western Australia, Crawley, WA 6009, Australia; mark.nicol@uwa.edu.au
[*] Correspondence: lisa.stinson@uwa.edu.au; Tel.: +61-8-6488-3200

† Presented at the Australian Breastfeeding + Lactation Research and Science Translation Conference (ABREAST Conference 2023), Perth, Australia, 10 November 2023.

Keywords: human milk; microbiota; associated factors

The human milk microbiota play an important role in mammary and infant health. However, there are limited data describing the composition and temporal variation in the milk microbiota beyond the first few months. Therefore, this study aimed to characterise the development of the human milk microbiota over the first two years postpartum and to investigate the factors that influence its composition. We performed full-length 16S rRNA gene sequencing on 608 samples collected from 86 mothers in the BLOSOM birth cohort. Similar to other populations, the milk microbiota of Western Australian women was low in diversity, and strongly dominated by *Staphylococcus* and *Streptococcus* species (14.95% and 39.04% relative abundance, respectively). Milk bacterial richness increased over time ($p = 0.004$), and there were significant changes in the abundances of 7 of 18 of the most abundant taxa over time. The composition of the milk microbiota was associated with numerous maternal and infant factors including infant age, maternal or infant antibiotic exposure, maternal probiotic use, parity, mode of delivery, and pre-pregnancy BMI (all $p < 0.039$). Our data provide new insights into how the human milk microbiota develop beyond the first few months and determinants of the milk's bacterial composition, demonstrating that the milk microbiota are dynamic over time in the first two years postpartum.

Author Contributions: Conceptualisation, L.F.S., M.P.N. and D.T.G.; methodology, L.F.S.; formal analysis, R.X. and L.F.S.; investigation, R.X. and A.S.C.; resources, D.T.G.; data curation, R.X.; writing—original draft preparation, R.X.; writing—review and editing, L.F.S. and D.T.G.; visualisation, R.X.; supervision, L.F.S., D.T.G. and M.P.N.; funding acquisition, D.T.G. All authors have read and agreed to the published version of the manuscript.

Funding: This research is funded by an unrestricted research grant from Medela AG (Switzerland).

Institutional Review Board Statement: The study was conducted in accordance with the Declaration of Helsinki and approved by the Human Research Ethics Committee of The University of Western Australia (RA/4/20/4023).

Informed Consent Statement: Informed consent was obtained from all subjects involved in this study.

Data Availability Statement: Sequence data have been submitted to the NCBI SRA (BioProject Submission: SUB13951443).

Citation: Xu, R.; Cheema, A.S.; Nicol, M.P.; Geddes, D.T.; Stinson, L.F. The Development of the Human Milk Microbiota over the First Two Years Postpartum in the Breastfeeding Longitudinal Observational Study of Mothers and Kids (BLOSOM) Cohort. *Proceedings* **2023**, *93*, 3. https://doi.org/10.3390/proceedings2023093003

Published: 19 December 2023

Acknowledgments: The authors would like to acknowledge Erika van der Dries for collecting the samples and Matthew Payne for the use of his laboratory space and equipment.

Conflicts of Interest: D.T.G. declares their participation in the Scientific Advisory Board of Medela. AG. D.T.G. and L.F.S. receive funding from Medela AG administered through The University of Western Australia. The funders had no role in the design of the study; in the collection, analyses, or interpretation of data; in the writing of the manuscript; or in the decision to publish the results.

proceedings

MDPI

Abstract

Microbial Underpinnings of Mastitis: Current State of the Evidence †

Lisa F. Stinson [1,2,3,*] and Donna T. Geddes [1,2,3]

1 School of Molecular Sciences, The University Western Australia, Crawley, WA 6009, Australia; donna.geddes@uwa.edu.au
2 ABREAST Network, Perth, WA 6000, Australia
3 UWA Centre for Human Lactation Research and Translation, Crawley, WA 6009, Australia
* Correspondence: lisa.stinson@uwa.edu.au
† Presented at the Australian Breastfeeding + Lactation Research and Science Translation Conference (ABREAST Conference 2023), Perth, Australia, 10 November 2023.

Abstract: Mastitis is an inflammatory condition of the breast, with or without accompanying infection. The recent release of the revised Academy of Breastfeeding Medicine protocol on mastitis (Clinical Protocol #36) has caused controversy within the field. The updated protocol positions multiple typical human milk commensal bacteria as causative agents of mastitis. However, data to support these relationships are lacking. Here, we critically review human and animal evidence for the role of the milk microbiota in mastitis. Only three metataxonomic/metagenomic studies have characterized the human milk microbiome in cases of mastitis. These studies consistently report reduced alpha diversity and elevated levels of *Staphylococcus aureus* in mastitic milk samples; however, the remaining findings presented are conflicting and inconsistent. Collectively, the three studies included 45 sub-acute mastitis cases, 24 acute mastitis cases, and 52 healthy controls, and are thus limited by low participant numbers. In addition, the studies vary in their definition of sub-acute/acute mastitis, their methodologies, and antibiotic exposure in the mastitic groups. Further, these studies provide data on the state of the microbiome during mastitis, with no data currently available on the milk microbiome preceding the onset of mastitis. These kind of longitudinal data are critical to identify candidates for disease causation. Emerging evidence from animal models is suggestive of the involvement of the gut microbiota. Studies have reported that fecal microbiota transplantation from mastitic cows to germ-free mice results in mastitis symptoms. Future studies should therefore consider the maternal microbiome more broadly when assessing the etiology of mastitis. While *S. aureus* is frequently recognized as a mastitis-related pathogen, data from culture-based and culture-independent studies demonstrate that this species is present in healthy women and cannot be detected in a significant portion of cases. This suggests heterogenous causes for bacterial mastitis, necessitating broader screening. Overall, data in this field are sparse, and current clinical guidelines lack high-quality evidence to support them. There is therefore a pressing need for further research in this area to better characterize the causes of mastitis and thereby underpin future therapeutics.

Keywords: mastitis; microbiome; breast inflammation; breast infection

Citation: Stinson, L.F.; Geddes, D.T. Microbial Underpinnings of Mastitis: Current State of the Evidence. *Proceedings* **2023**, *93*, 4. https://doi.org/10.3390/proceedings2023093004

Academic Editors: Debra J. Palmer and Nicolas L. Taylor

Published: 19 December 2023

Author Contributions: Conceptualization, L.F.S. and D.T.G.; investigation, L.F.S.; writing—original draft preparation, L.F.S.; writing—review and editing, D.T.G. All authors have read and agreed to the published version of the manuscript.

Funding: This research received no external funding.

Institutional Review Board Statement: Not applicable.

Informed Consent Statement: Not applicable.

Data Availability Statement: Not applicable.

Conflicts of Interest: D.T.G. declares participation in the Scientific Advisory Board of Medela AG. L.F.S. and D.T.G. are/were supported by an unrestricted research grant from Medela AG, administered by The University of Western Australia. The funders had no role in the design of the review; in the collection, analyses, or interpretation of data; in the writing of the manuscript; or in the decision to publish the results.

proceedings

MDPI

Abstract

Biomarkers of Low Milk Supply †

Ching Tat Lai [1,2,3,*], Xuehua Jin [1,2,3], Xiaojie Zhou [1,2,3], Lisa F. Stinson [1,2,3], Mary E. Wlodek [1,4], Sharon L. Perrella [1,2,3], Zoya Gridneva [1,2,3] and Donna T. Geddes [1,2,3]

1 School of Molecular Sciences, The University of Western Australia, Crawley, WA 6009, Australia; xuehua.jin@research.uwa.edu.au (X.J.); xiaojie.zhou@uwa.edu.au (X.Z.); lisa.stinson@uwa.edu.au (L.F.S.); m.wlodek@unimelb.edu.au (M.E.W.); sharon.perrella@uwa.edu.au (S.L.P.); zoya.gridneva@uwa.edu.au (Z.G.); donna.geddes@uwa.edu.au (D.T.G.)
2 ABREAST Network, Perth, WA 6000, Australia
3 UWA Centre for Human Lactation Research and Translation, Crawley, WA 6009, Australia
4 Department of Obstetrics and Gynaecology, The University of Melbourne, Melbourne, VIC 3010, Australia
* Correspondence: ching-tat.lai@uwa.edu.au; Tel.: +61-8-6488-3200

† Presented at the Australian Breastfeeding + Lactation Research and Science Translation Conference (ABREAST Conference 2023), Perth, Australia, 10 November 2023.

Keywords: low milk supply; biomarkers; milk component; predictive model

It is estimated that 5–10% of breastfeeding mothers are unable to produce enough milk for their infants. Causes of low milk supply are multifactorial, including factors such as gene mutations, endocrine disorders and infrequent milk removal. All these factors affect the normal function of the mammary gland and potentially the concentrations of human milk components. Therefore, an array of milk components may act as biomarkers for the detection of low milk supply. Various biochemical assays were used to measure an array of milk components (fat, total protein, casein, whey, lactose, citrate, calcium, copper, iron, potassium, magnesium, sodium, phosphorous and zinc) in 48 women with low milk supply (<600 mL/24 h) and 65 with normal milk supply (>600 mL/24 h). Univariable linear regression and correlation analysis were used to examine associations between milk component concentrations and milk supply. The dataset was then randomly split into a training set (70%) and a test set (30%). Multivariable logistic regression with the training set was used to develop prediction models with various combinations of milk component concentrations, and the test set was used to validate the predictivity of these models.

Univariable linear regression showed that concentrations of fat ($p = 0.039$), total protein ($p = 0.002$), whey ($p = 0.001$), potassium ($p = 0.02$), sodium ($p = 0.001$) and zinc ($p = 0.001$) and the Na:K ratio ($p = 0.008$) were significantly different between samples from low- and normal-milk-supply mothers. In addition, total protein, casein, whey, lactose, citrate, calcium, potassium, sodium, phosphorous, zinc and the Na:K ratio were strongly correlated with each other ($r > 0.6$ or $r < -0.6$). The predictive model with fat content as the sole predictor had an accuracy of 61%, precision of 80% and recall of 52%, while the model with fat, total protein and lactose as predictors had an accuracy of 53%, precision of 100% and recall of 33% in predicting low milk supply. Further, the model with only sodium and potassium had an accuracy of 64%, precision of 80% and recall of 50%. The best predictive model included all measured milk components, and had an accuracy of 79%, precision of 84% and recall of 76% in predicting low milk supply.

These preliminary findings suggest that differences in milk composition between women with low and normal milk supply have the potential to be used in testing for the detection of low milk production.

Citation: Lai, C.T.; Jin, X.; Zhou, X.; Stinson, L.F.; Wlodek, M.E.; Perrella, S.L.; Gridneva, Z.; Geddes, D.T. Biomarkers of Low Milk Supply. *Proceedings* **2023**, *93*, 5. https://doi.org/10.3390/proceedings2023093005

Academic Editors: Debra J. Palmer and Nicolas L. Taylor

Published: 19 December 2023

Author Contributions: Conceptualization, C.T.L. and D.T.G.; methodology, D.T.G., C.T.L., X.J., X.Z., L.F.S., S.L.P. and Z.G.; data collection, C.T.L., X.J., X.Z., S.L.P. and Z.G.; formal analysis, C.T.L., X.J. and X.Z.; investigation, D.T.G. and C.T.L.; resources, D.T.G.; data curation, D.T.G., C.T.L. and X.J.; writing—original draft preparation, C.T.L.; writing—review and editing, D.T.G., X.J., X.Z., L.F.S., M.E.W., S.L.P. and Z.G.; visualization, C.T.L.; supervision, D.T.G.; funding acquisition, D.T.G. All authors have read and agreed to the published version of the manuscript.

Funding: This research was funded by an unrestricted research grant from Medela AG (Switzerland). C.T.L., X.Z., L.F.S., S.L.P., Z.G. and D.T.G. receive a salary from an unrestricted research grant paid by Medela AG and administered by The University of Western Australia. X.J. is supported by a UWA—CSC Joint PhD Scholarship and a UWA—CSC Higher Degree by Research Top-Up Scholarship.

Institutional Review Board Statement: This study was conducted in accordance with the Declaration of Helsinki. The study was approved by the Human Research Ethics Committee at The University of Western Australia (2019/RA/4/20/4023) and conducted in accordance with the relevant guidelines and regulations.

Informed Consent Statement: Informed consent was obtained from all subjects involved in this study.

Data Availability Statement: Data sharing is not applicable.

Acknowledgments: We thank all our participants and their families for their time and help with this research.

Conflicts of Interest: D.T.G. declares participation in the Scientific Advisory Board of Medela AG. C.T.L., X.Z., L.F.S., S.L.P., Z.G. and D.T.G. are/were supported by an unrestricted research grant from Medela AG, administered by The University of Western Australia. The funders had no role in the design of the study; in the collection, analyses, or interpretation of the data; in the writing of the manuscript; or in the decision to publish the results. All other authors declare no conflicts of interest.

proceedings

MDPI

Abstract

Are Low Lactose Concentrations a Risk Factor for *Staphylococcus aureus*-Associated Mastitis? †

Grace C. McLoughlin [1,2,3,*], Stefano Cacciatore [4], Anna O. Okunola [5], Noor-Ul-Huda Ghori [1], Heather J. Zar [6], Donna T. Geddes [2,3,7] and Mark P. Nicol [1]

1 School of Biomedical Sciences, The University of Western Australia, Crawley, WA 6009, Australia; noor-ul-huda.ghori@uwa.edu.au (N.-U.-H.G.); mark.nicol@uwa.edu.au (M.P.N.)
2 ABREAST Network, Perth, WA 6000, Australia; donna.geddes@uwa.edu.au
3 UWA Centre for Human Lactation Research and Translation, Crawley, WA 6009, Australia
4 International Centre for Genetic Engineering and Biotechnology (ICGEB), Cape Town 7925, South Africa; stefano.cacciatore@icgeb.org
5 Division of Molecular Biology and Human Genetics, Faculty of Medicine and Health Sciences, DSI-NRF Centre of Excellence for Biomedical Tuberculosis Research, South African Medical Research Council Centre for Tuberculosis Research, Stellenbosch University, Cape Town 7505, South Africa; annaojo@sun.ac.za
6 Department of Pediatrics and Child Health, Red Cross Memorial Children's Hospital, Cape Town 7925, South Africa; heather.zar@uct.ac.za
7 School of Molecular Sciences, The University of Western Australia, Crawley, WA 6009, Australia
* Correspondence: grace.mcloughlin@research.uwa.edu.au
† Presented at the Australian Breastfeeding + Lactation Research and Science Translation Conference (ABREAST Conference 2023), Perth, Australia, 10 November 2023.

Abstract: *Staphylococcus aureus* is a bacterium found in the milk of up to 38% of healthy lactating mothers; however, *S. aureus* is isolated with increased frequency from colostrum and mastitis milk. Both of these milk types have lower lactose concentrations compared to mature milk from healthy lactating mothers, which may indicate that lactose has a role in determining whether *S. aureus* can survive in human milk. The aim of this study was (1) to investigate whether the presence of *S. aureus* in human milk is associated with the milk's lactose concentration, and (2) to determine whether different lactose concentrations can affect the ability of *S. aureus* isolates to grow in vitro. Human milk samples were collected at 10 weeks postpartum from mothers participating in the Drakenstein Child Health Study (Cape Town, South Africa) and underwent NMR spectroscopy to determine their metabolome. A subset of these samples ($n = 117$) was cultured to isolate *S. aureus*. Milk samples with lactose concentrations of less than 166 mM were more likely to have *S. aureus* present, compared to samples with lactose concentrations of over 166 mM ($p < 0.001$). In vitro, the growth of *S. aureus* was negatively correlated with the lactose concentration of axenic culture. Lactose concentrations associated with human milk appear to have an inhibitory effect on the growth of *S. aureus* human milk isolates. Therefore, low-lactose human milk could potentially be a risk factor for increased *S. aureus* growth and the development of *S. aureus*-associated mastitis.

Keywords: *Staphylococcus aureus*; human milk; lactose; mastitis; culture

Citation: McLoughlin, G.C.; Cacciatore, S.; Okunola, A.O.; Ghori, N.-U.-H.; Zar, H.J.; Geddes, D.T.; Nicol, M.P. Are Low Lactose Concentrations a Risk Factor for *Staphylococcus aureus*-Associated Mastitis?. *Proceedings* **2023**, *93*, 6. https://doi.org/10.3390/proceedings2023093006

Academic Editors: Debra J. Palmer and Nicolas L. Taylor

Published: 19 December 2023

Author Contributions: Conceptualization, G.C.M., N.-U.-H.G. and M.P.N.; methodology, G.C.M., N.-U.-H.G., A.O.O., S.C., H.J.Z., D.T.G. and M.P.N.; software, G.C.M.; formal analysis, G.C.M., N.-U.-H.G., D.T.G. and M.P.N.; investigation, G.C.M., D.T.G. and M.P.N.; resources, H.J.Z. and M.P.N.; data curation, G.C.M., A.O.O. and S.C.; writing—original draft preparation, G.C.M.; writing—review and editing, G.C.M., N.-U.-H.G., D.T.G. and M.P.N.; supervision, N.-U.-H.G., D.T.G. and M.P.N.; funding acquisition, H.J.Z. and M.P.N. All authors have read and agreed to the published version of the manuscript.

Funding: This work was supported by the Bill and Melinda Gates Foundation (OPP1017641; OPP1017579), and the Australian National Health and Medical Research Council (APP1174455). G.M. is supported by the Sibyl and Wylie Gibbons Scholarship award from the Marshall Centre (The University of Western Australia). D.T.G. is supported by an unrestricted research grant from Medela AG (Switzerland), administered by The University of Western Australia. The funders had no role in the design of the study; in the collection, analyses, or interpretation of data; in the writing of the manuscript; or in the decision to publish the results.

Institutional Review Board Statement: The study was conducted in accordance with the Declaration of Helsinki, and approved by the Human Research Ethics Committee of the University of Western Australia (protocol code 2019/RA/5/15/1294 and date of approval 27 May 2020).

Informed Consent Statement: Informed consent was obtained from all subjects involved in the study.

Data Availability Statement: Data sharing is not applicable.

Conflicts of Interest: D.T.G. declares participation in the Scientific Advisory Board of Medela AG. D.T.G. is supported by an unrestricted research grant from Medela AG, administered by The University of Western Australia. The funders had no role in the design of the study; in the collection, analyses, or interpretation of data; in the writing of the manuscript; or in the decision to publish the results. All other authors declare no conflict of interest.

proceedings

MDPI

Abstract

Unravelling the Determinants of the Human Milk Microbiome in Allergic Women †

Jie Ma [1,2,3], Debra J. Palmer [4,5], Ching Tat Lai [1,2,3], Donna T. Geddes [1,2,3] and Lisa F. Stinson [1,2,3,*]

1 School of Molecular Sciences, University of Western Australia, Crawley, WA 6009, Australia; jie.ma@research.uwa.edu.au (J.M.); ching-tat.lai@uwa.edu.au (C.T.L.); donna.geddes@uwa.edu.au (D.T.G.)
2 ABREAST Network, Perth, WA 6000, Australia
3 UWA Centre for Human Lactation Research and Translation, Crawley, WA 6009, Australia
4 Telethon Kids Institute, University of Western Australia, Nedlands, WA 6009, Australia; debbie.palmer@telethonkids.org.au
5 School of Medicine, University of Western Australia, Crawley, WA 6009, Australia
* Correspondence: lisa.stinson@uwa.edu.au

† Presented at the Australian Breastfeeding + Lactation Research and Science Translation Conference (ABREAST Conference 2023), Perth, Australia, 10 November 2023.

Keywords: human milk; maternal allergic disease; microbiome

As some of the most common chronic diseases globally, allergic diseases have been associated with the gut microbiome composition of both children and adults. The human milk microbiome has been proposed as a potential mediator in infant microbiome development; however, no studies have explored the human milk microbiome specifically in allergic women. This study aimed to investigate the determinants of the human milk microbiome in allergic women. Milk samples from the Infant Fish Oil Supplementation Study (IFOS; Australian Clinical Trials Registry ACTRN12606000281594) were collected at three (n = 229) and six (n = 171) months postpartum from women, who all had at least one allergic disease (e.g., asthma, allergic rhinitis). Full-length 16S rRNA gene sequencing was utilised to characterise the milk microbiome. Differences in alpha diversity and microbiome composition were assessed by linear mixed models, while beta diversity was assessed by PERMANOVA. The season of infant birth had the strongest relationship with the milk microbiome, with significant associations with beta diversity (p = 0.001) and 10 out of the 19 most abundant OTUs (>0.1% relative abundance). The milk microbiome composition was also significantly associated with the delivery mode, feeding status, parity, infant age, infant sex, maternal age, passive smoke exposure, and pets at home, aligning with previous findings from non-allergic women. Maternal allergy type was also significantly related to the composition of the milk microbiome. This study investigated the human milk microbiome in allergic mothers, revealing the season of infant birth as potentially the strongest influencer, along with associations between maternal, infant, and environmental factors. These findings suggest an interaction between season and maternal allergic diseases, which influences the microbiome.

Citation: Ma, J.; Palmer, D.J.; Lai, C.T.; Geddes, D.T.; Stinson, L.F. Unravelling the Determinants of the Human Milk Microbiome in Allergic Women. *Proceedings* **2023**, *93*, 7. https://doi.org/10.3390/proceedings2023093007

Academic Editors: Zoya Gridneva and Nicolas L. Taylor

Published: 19 December 2023

Author Contributions: Conceptualization, D.T.G. and L.F.S.; methodology, L.F.S.; data collection D.J.P.; formal analysis, J.M.; investigation, J.M.; resources, D.T.G.; data curation, D.J.P.; writing—original draft preparation, J.M.; writing—review and editing, L.F.S., D.T.G. and D.J.P.; visualization, J.M.; supervision, L.F.S., D.T.G., D.J.P. and C.T.L.; project administration, L.F.S.; funding acquisition, D.T.G. All authors have read and agreed to the published version of the manuscript.

Funding: J.M., L.F.S., C.T.L., and D.T.G. are supported by an unrestricted research grant from Medela AG (Switzerland), administered by The University of Western Australia. J.M. was supported by an additional SIRF (Scholarships for International Research Fees) scholarship from The University of Western Australia and Ad Hoc Postgraduate Scholarship from Medela AG. D.J.P. is supported by a Telethon Kids Institute Ascend Fellowship. The funding bodies had no role in the design of the study; in the collection, analyses, or interpretation of data; in the writing of the manuscript; or in the decision to publish the results.

Institutional Review Board Statement: The study was conducted in accordance with the Declaration of Helsinki and was approved by the Princess Margaret Hospital Human Research Ethics Committee (HREC approval number 1111EP).

Informed Consent Statement: Written informed consent was obtained from all participants involved in the study.

Data Availability Statement: Data sharing not applicable.

Conflicts of Interest: D.T.G. declares participation in the Scientific Advisory Board of Medela AG. J.M., L.F.S., C.T.L. and D.T.G. are supported by an unrestricted research grant from Medela AG, administered by The University of Western Australia. The funders had no role in the design of the study; in the collection, analyses, or interpretation of data; in the writing of the manuscript; or in the decision to publish the results. D.J.P. declares no conflict of interest.

proceedings

Abstract

Can Arginine Help to Improve Milk Supply in Humans? It Does in Cows †

Luoyang Ding [1,2,*], Yizhao Shen [1], Tianyou Wu [1], Lianmin Chen [1], Juan J. Loor [3], Shane K. Maloney [4], Mengzhi Wang [1] and Dominique Blache [1,2]

1 College of Animal Science and Technology, Yangzhou University, Yangzhou 225009, China; shenyizhao@163.com (Y.S.); 18752540287@163.com (T.W.); lianminchen@yeah.net (L.C.); mzwang@yzu.edu.cn (M.W.); dominique.blache@uwa.edu.au (D.B.)
2 School of Agriculture and Environment, The University of Western Australia, Crawley, WA 6009, Australia
3 Department of Animal Sciences and Division of Nutritional Sciences, University of Illinois, Urbana, IL 61801, USA; jloor@illinois.edu
4 School of Human Sciences, The University of Western Australia, Crawley, WA 6009, Australia; shane.maloney@uwa.edu.au
* Correspondence: luoyang.ding@uwa.edu.au
† Presented at the Australian Breastfeeding + Lactation Research and Science Translation Conference (ABREAST Conference 2023), Perth, Australia, 10 November 2023.

Keywords: arginine; milk production; nutrients utilization efficiency; AMP-activated protein kinase

Arginine can be metabolized into nitric oxide, polyamine, creatine, or agmatine, and each of those metabolites has several biological functions. Arginine has been shown to play an important role in the regulation of metabolism, immune function, and hormone secretion in mammals. The present study tested the hypothesis that arginine could be beneficial to milk supply by enhancing the efficiency of nutrient utilization. Using lactating cows as the study model, we found that supplementation with arginine via a jugular vein increased the daily milk yield, milk protein yield, and milk fat yield. The supplementation had no effect on the feed intake or the digestibility of dry matter, crude protein, or ether extract. The urea nitrogen in serum, urine, and milk was lower in cows that were infused with arginine, indicating a better utilization of nitrogen in the cows that were supplied with extra arginine. The underlying cause of these changes may have been the elevated serum nitric oxide (a potent vasorelaxant in mammals), stimulating mammary blood flow and the supply of amino acids, fatty acids (FAs), and glucose to the mammary gland in the cows that were infused with arginine. Furthermore, the expression of genes that code for amino acid transporters (*SLC7A2* and *SLC7A8*), and enzymes involved in the biosynthesis of FA (*ACACA*) and triglycerides (*SCD*), and FA desaturation (*DGAT1*) were higher in the mammary gland of cows that were infused with arginine. As a result, the de novo synthesis of FA and casein in the mammary gland were enhanced in the cows that were supplied with extra arginine. Taken together, the available data suggest that the positive effect of arginine on milk supply was caused by a combination of more blood flow via nitric oxide and a direct effect of arginine on the expression of genes that code for proteins that are involved in the synthesis of milk protein and milk fat.

Citation: Ding, L.; Shen, Y.; Wu, T.; Chen, L.; Loor, J.J.; Maloney, S.K.; Wang, M.; Blache, D. Can Arginine Help to Improve Milk Supply in Humans? It Does in Cows. *Proceedings* **2023**, *93*, 8. https://doi.org/10.3390/proceedings2023093008

Academic Editors: Donna T. Geddes and Zoya Gridneva

Published: 20 December 2023

Author Contributions: Conceptualization, L.D., M.W. and D.B.; methodology, L.D., Y.S., T.W. and L.C.; formal analysis, L.D., D.B. and S.K.M.; investigation, L.D., Y.S. and T.W.; resources, J.J.L.; writing—original draft preparation, L.D.; writing—review and editing, M.W., D.B. and S.K.M.; supervision, M.W. and D.B.; project administration, L.D.; funding acquisition, M.W. All authors have read and agreed to the published version of the manuscript.

Funding: This study was funded by the National Key Research and Development Program of China (2018YFD0502100) and the China Scholarship Council—The University of Western Australia Joint

Scholarship (201708320259). The funders had no role in the design of the study; in the collection, analyses, or interpretation of data; in the writing of the manuscript; or in the decision to publish the results.

Institutional Review Board Statement: All the procedures for the treatment and care of experimental animals were approved by the Yangzhou University Animal Care and Use Committee (EA/201900011) and followed the established guidelines for animal welfare.

Informed Consent Statement: Not applicable.

Data Availability Statement: The data analyzed during the current study are available from the corresponding author on reasonable request.

Acknowledgments: The authors of this manuscript thank the staff at Experimental Farm of Yangzhou University (Yangzhou, Jiangsu, China) for their support in taking care of the animals.

Conflicts of Interest: The authors declare no conflicts of interest.

proceedings

MDPI

Abstract

Human Milk Expression Technologies: An Evaluation of Mobility and Comfort Perception of a Hands-Free, In-Bra, Breastmilk Collection Pump Set †

Zoya Gridneva *, Ashleigh H. Warden, Jacki L. McEachran, Ching Tat Lai, Sharon L. Perrella and Donna T. Geddes

School of Molecular Sciences, The University of Western Australia, Crawley, WA 6009, Australia; ashleigh.warden@uwa.edu.au (A.H.W.); jacki.mceachran@uwa.edu.au (J.L.M.); ching-tat.lai@uwa.edu.au (C.T.L.); sharon.perrella@uwa.edu.au (S.L.P.); donna.geddes@uwa.edu.au (D.T.G.)

* Correspondence: zoya.gridneva@uwa.edu.au; Tel.: +61-8-6488-4467

† Presented at the Australian Breastfeeding + Lactation Research and Science Translation Conference (ABREAST Conference 2023), Perth, Australia, 10 November 2023.

Keywords: lactation; human milk; breast pumping; breast expression; electric pump; comfortable pumping; hands-free pump; wearable pump

Citation: Gridneva, Z.; Warden, A.H.; McEachran, J.L.; Lai, C.T.; Perrella, S.L.; Geddes, D.T. Human Milk Expression Technologies: An Evaluation of Mobility and Comfort Perception of a Hands-Free, In-Bra, Breastmilk Collection Pump Set. *Proceedings* **2023**, *93*, 9. https://doi.org/10.3390/proceedings2023093009

Academic Editors: Debra J. Palmer and Nicolas L. Taylor

Published: 21 December 2023

Wearable pumps have been designed to improve convenience and maximize flexibility whilst pumping and are reported to benefit lactating health professionals when they return to work [1]. However, their performance in regard to efficacy, comfort and ease of use is rarely evaluated. One recent study reported on the efficacy and comfort of an experimental hands-free, in-bra, breastmilk collection pump set (IBCPS) connected to a personal use double electric breast pump [2], now available as a Freestyle Hands-Free Breast Pump. Applying the IBCPS resulted in efficient and effective breast emptying with good comfort ratings. We present a further evaluation of its mobility, convenience, and ease of use as rated by participants. Twenty-three lactating mothers 1–6 months postpartum participated in this cross-sectional study. Mothers completed one pumping session with IBCPS where both breasts were pumped simultaneously using the participant's maximum comfortable vacuum for a period of 15 min of expression after the first milk ejection. Mobility testing was conducted throughout a series of positions (whilst walking, reaching up with arms stretched, and leaning over) during the expression phase of pumping, and participant satisfaction and confidence, and comparison to home pump questionnaires were completed. A scale from 1 to 5 was used to rate the participants' degrees of comfort, confidence, satisfaction and ease of use, general pumping experience and likelihood of multitasking when using the pump. A rating of 1 indicated very comfortable, confident, satisfied, very easy to use, very good and very likely; and 5 indicated very uncomfortable, absolutely not confident, very unsatisfied, very difficult to use, very bad and very unlikely. The results are presented as mean and standard deviation values and minimum–maximum values, which generally were between 1 and 3, unless specifically reported. During the pumping session, the mean initial, mid pumping and final comfort levels were 1.9 ± 0.6, 1.8 ± 0.7 and 1.9 ± 0.8, respectively [2]. Mothers' confidence in pumping while walking was 1.4 ± 0.6, 1.5 ± 0.7 for reaching up, and 1.9 ± 1.0 (1–4) for leaning over. Satisfaction with the amount of expressed milk was 1.9 ± 1.2, with comfort 1.4 ± 0.6, with collection pump set fit in the bra 1.7 ± 0.7, with the bra shape with pump set in it 1.5 ± 0.8, with pump set weight in the bra 1.0 ± 0.2 (1–2) and with applied vacuum 1.4 ± 0.7. Mothers found the IBCPS easy to use, to connect and place in the bra (1.9 ± 0.9 (1–4)), to align the nipple (1.7 ± 0.8), to remove the pump set from the bra (1.6 ± 0.7) and to pour milk from pump set into the bottle (1.4 ± 0.6). The mothers rated overall pumping experiences with IBCPS as very good (1.4 ± 0.5 (1–2) and indicated that they would be very likely to multitask when using the IBCPS (1.4 ± 0.6). When comparing the IBCPS to their home pumps, the mothers liked IBCPS more (1.4 ± 0.7) and rated it as more effective (2.2 ± 1.1 (1–5)), more comfortable (2.0 ± 0.9 (1–4)) and generally better than their home pumps (2.1 ± 1.0 (1–4)). Overall, the mothers rated pumping experience with IBCPS as satisfactory, comfortable or better and they found the pump easy to use and liked the hands-free aspect and the mobility. These findings further support the use of wearable pumps to allow for maximal flexibility for mothers that wish to combine their lactation and career goals.

Author Contributions: Conceptualization, Z.G. and D.T.G.; methodology, D.T.G., C.T.L. and Z.G.; data collection, A.H.W., D.T.G. and Z.G.; formal analysis, Z.G. and A.H.W.; investigation, A.H.W., D.T.G. and Z.G.; resources, D.T.G.; data curation, A.H.W., D.T.G., J.L.M. and Z.G.; writing—original draft preparation, Z.G.; writing—review and editing, A.H.W., J.L.M., C.T.L., S.L.P. and D.T.G.; visualization, Z.G.; supervision, D.T.G.; project administration, J.L.M.; funding acquisition, D.T.G. All authors have read and agreed to the published version of the manuscript.

Funding: This research was funded by an unrestricted research grant from Medela AG (Switzerland). The funder had no role in the design of the study; in the collection, analyses, or interpretation of data; in the writing of the manuscript; or in the decision to publish the results.

Institutional Review Board Statement: The study was conducted in accordance with the Declaration of Helsinki. The study was approved by the Human Research Ethics Committee at The University of Western Australia (RA/4/20/6407) and conducted in accordance with the relevant guidelines and regulations.

Informed Consent Statement: Informed consent was obtained from all subjects involved in the study.

Data Availability Statement: Restrictions apply to the availability of some, or all data generated or analyzed during this study. The corresponding author will on request detail the restrictions and any conditions under which access to some data may be provided.

Acknowledgments: We thank all our participants and their families for their time and help with this research.

Conflicts of Interest: D.T.G. declares participation in the Scientific Advisory Board of Medela AG. Z.G., A.H.W., J.L.M., C.T.L., S.L.P and D.T.G. are/were supported by an unrestricted research grant from Medela AG, administered by The University of Western Australia. The funders had no role in the design of the study; in the collection, analyses, or interpretation of data; in the writing of the manuscript; or in the decision to publish the results.

References

1. Colbenson, G.A.; Hoff, O.C.; Olson, E.M.; Ducharme-Smith, A. The impact of wearable breast pumps on physicians' breastfeeding experience and success. *Breastfed. Med.* **2022**, *17*, 537–543. [CrossRef] [PubMed]
2. Gridneva, Z.; Warden, A.H.; McEachran, J.L.; Lai, C.T.; Perrella, S.L.; Geddes, D.T. Human milk expression technologies: An evaluation of efficacy and comfort of hands-free, in-bra, breastmilk collection pump set. *Clin. Nutr. Open Sci.* **2023**, *49*, 28–36. [CrossRef]

proceedings

MDPI

Abstract

Breastfeeding Duration and Bone Mineral Density in Childhood: A Prospective Study within GUSTO Cohort †

Zoya Gridneva [1,2,3,4,*], Wei Wei Pang [5,6], Philip Vlaskovsky [7], Jacki L. McEachran [1,3,4], Sharon L. Perrella [1,2,3,4], Fabian Yap [8], Mary E. Wlodek [1,9,10,11,12], Yap-Seng Chong [10,11,12], Johan G. Eriksson [12,13,14,15], Donna T. Geddes [1,2,3,4] and Mya Thway Tint [12,13]

1 School of Molecular Sciences, The University of Western Australia, Crawley, WA 6009, Australia; jacki.mceachran@uwa.edu.au (J.L.M.); sharon.perrella@uwa.edu.au (S.L.P.); m.wlodek@unimelb.edu.au (M.E.W.); donna.geddes@uwa.edu.au (D.T.G.)
2 International Society for Research in Human Milk and Lactation, Minneapolis, MN 55416, USA
3 ABREAST Network, Perth, WA 6000, Australia
4 UWA Centre for Human Lactation Research and Translation, Crawley, WA 6009, Australia
5 Global Center for Asian Women's Health (GloW), Yong Loo Lin School of Medicine, National University of Singapore, Singapore 117597, Singapore; obgpww@nus.edu.sg
6 Bia-Echo Asia Centre for Reproductive Longevity & Equality (ACRLE), Yong Loo Lin School of Medicine, National University of Singapore, Singapore 117597, Singapore
7 Department of Mathematics and Statistics, School of Physics, Mathematics and Computing, The University of Western Australia, Crawley, WA 6009, Australia; philip.vlaskovsky@uwa.edu.au
8 Department of Paediatrics, KK Women's and Children's Hospital, Singapore 229899, Singapore; fabian.yap.k.p@singhealth.com.sg
9 Department of Obstetrics and Gynaecology, The University of Melbourne, Melbourne, VIC 3010, Australia
10 Department of Obstetrics and Gynaecology, Yong Loo Lin School of Medicine, National University of Singapore, Singapore 117597, Singapore; obgcys@nus.edu.sg
11 National University Health System, Singapore 119228, Singapore
12 Singapore Institute for Clinical Sciences (SICS), Agency for Science, Technology and Research (A*STAR), Singapore 117609, Singapore; obgjge@nus.edu.sg (J.G.E.); mya_thway_tint@sics.a-star.edu.sg (M.T.T.)
13 Human Potential Translational Research Programme, Yong Loo Lin School of Medicine, National University of Singapore, Singapore 117597, Singapore
14 Department of General Practice and Primary Health Care, University of Helsinki, 00014 Helsinki, Finland
15 Folkhälsan Research Centre, 00250 Helsinki, Finland
* Correspondence: zoya.gridneva@uwa.edu.au; Tel.: +61-8-6488-4467

† Presented at the Australian Breastfeeding + Lactation Research and Science Translation Conference (ABREAST Conference 2023), Perth, Australia, 10 November 2023.

Citation: Gridneva, Z.; Pang, W.W.; Vlaskovsky, P.; McEachran, J.L.; Perrella, S.L.; Yap, F.; Wlodek, M.E.; Chong, Y.-S.; Eriksson, J.G.; Geddes, D.T.; et al. Breastfeeding Duration and Bone Mineral Density in Childhood: A Prospective Study within GUSTO Cohort. *Proceedings* **2023**, *93*, 10. https://doi.org/10.3390/proceedings2023093010

Published: 22 December 2023

Keywords: lactation; breastfeeding duration; infants and children; bone mineral density; dual-energy X-ray absorptiometry; lumbar spine; solid food introduction; growth; GUSTO cohort; Asian

Nutrition contributes to bone mineral density (BMD) and plays a role in bone growth during infancy and childhood. However, the published relationships between breastfeeding exposure and BMD in infancy, childhood and adult life are not consistent. This cross-sectional study examined relationships of both breastfeeding duration and time of solid food introduction, with BMD in young Asian children from the Growing up in Singapore Towards healthy Outcomes (GUSTO) cohort.

Six-year-old children that were born healthy at term with available dual-energy X-ray absorptiometry (DXA) lumbar spine (LS) scans and data on the duration of any breastfeeding were included in the analysis (n = 207; 103 boys, 104 girls; 110 Chinese, 34 Indian and 63 Malay). LS bone mineral apparent density (BMAD), i.e., volumetric BMD, was estimated according to the published equation based on bone mineral content (BMC) and bone area from L2 to L4 ($BMC/A_p{}^{3/2}$) [1]. Outcomes in univariable and multivariable linear regression models included areal BMD_{LS} ($aBMD_{LS}$) and $BMAD_{LS}$ (to take account of bone size in growing children), and the standard deviation scores $Z_{LS\text{-}BMAD}$ and $Z_{LS\text{-}aBMD}$.

Covariates adjusted for were maternal ethnicity, pre-pregnancy BMI, child's sex and both, child weight, and physical activity at 6 years of age.

Only 11 children were not breastfed, with the rest breastfed on average for 7.6 $\pm$ 11.5 (0.0–54.8) months. At 6 years of age, no significant difference in BMD_{LS} was detected by maternal ethnicity. Boys had lower BMD_{LS} compared to girls ($aBMD_{LS}$: -0.025 (95% CI: -0.040, -0.010; $p = 0.002$) g/cm^2; $Z_{LS\text{-}aBMD}$: -0.436 (95% CI: -0.704, -0.168; $p = 0.002$) g/cm^2; $BMAD_{LS}$: -0.008 (95% CI: -0.011, -0.005; $p < 0.0001$) g/cm^3; $Z_{LS\text{-}BMAD}$: -0.621 (95% CI: -0.882, -0.360; $p < 0.0001$) g/cm^3). In the univariable model, children with longer breastfeeding duration had significantly lower $aBMD_{LS}$, but not $BMAD_{LS}$ ($aBMD_{LS}$: -0.0007 (95% CI: -0.001, -1.575; $p = 0.045$) g/cm^2; $Z_{LS\text{-}aBMD}$: -0.012 (95% CI: -0.024, -0.0003; $p = 0.045$) g/cm^2). In the analysis stratified by sex, the association with breastfeeding duration was significant only for girls and only for $aBMD_{LS}$ ($aBMD_{LS}$: -0.001 (95% CI: -0.002, -0.0001; $p = 0.037$) g/cm^2; $Z_{LS\text{-}aBMD}$: -0.018 (95% CI: -0.035, -0.001; $p = 0.037$) g/cm^2). Adjusting for covariates resulted in no significant relationships with breastfeeding duration.

The average time of solid food introduction was reported as 5.5 $\pm$ 1.2 months (3.0–10.0; $n = 188$; 93 boys, 95 girls). No significant relationship with time of solid food introduction was found in the entire group in both the univariable and adjusted models. However, in stratified analysis, boys with later solid food introduction had lower BMD_{LS} ($aBMD_{LS}$: -0.009 (95% CI: -0.018, -0.0006; $p = 0.037$) g/cm^2; $Z_{LS\text{-}aBMD}$: -0.166 (95% CI: -0.321, -0.010; $p = 0.037$) g/cm^2; $BMAD_{LS}$: -0.003 (95% CI: -0.005, -0.0006; $p = 0.011$) g/cm^3; $Z_{LS\text{-}BMAD}$: -0.205 (95% CI: -0.360, -0.050; $p = 0.010$) g/cm^3). These relationships persisted when adjusting for covariates.

Breastfeeding duration and time of solid food introduction may have a long-term impact on bone mineralization in young children. However, a more precise quantitative approach when measuring breastfeeding exposure, such as 24 h infant milk intake and intake of milk components, would be prudent. Understanding the factors influencing bone remodeling during these periods of rapid skeletal growth is important for determining effective interventions to enhance bone development in vulnerable infants and children.

Author Contributions: Conceptualization of GUSTO, Y.-S.C. and F.Y.; conceptualization of this study, M.T.T., Z.G. and D.T.G.; methodology, M.T.T., W.W.P., P.V. and Z.G.; data collection, M.T.T. and W.W.P.; formal analysis, Z.G.; investigation, M.T.T., Z.G. and W.W.P.; resources, M.T.T.; data curation, M.T.T., Z.G. and W.W.P.; writing—original draft preparation, Z.G.; writing—review and editing, W.W.P., P.V., J.L.M., S.L.P., F.Y., M.E.W., Y.-S.C., J.G.E., D.T.G. and M.T.T.; visualization, Z.G.; supervision, M.T.T. and D.T.G.; project administration, J.L.M.; funding acquisition, Y.-S.C., D.T.G. and J.L.M. All authors have read and agreed to the published version of the manuscript.

Funding: The GUSTO study is supported by the National Research Foundation (NRF) under the Open Fund-Large Collaborative Grant (OF-LCG; MOH-000504) administered by the Singapore Ministry of Health's National Medical Research Council (NMRC) and the Agency for Science, Technology and Research (A*STAR). Z.G. was funded by an unrestricted research grant from Medela AG (Switzerland) and has been financially supported (in part) by the International Society for Research in Human Milk and Lactation - Family Larsson-Rosenquist Foundation Trainee Expansion Program. The funders had no role in the design of the study; in the collection, analyses, or interpretation of data; in the writing of the manuscript; or in the decision to publish the results.

Institutional Review Board Statement: This study was conducted in accordance with the Declaration of Helsinki and was approved by the Institutional Review Board of the National Healthcare Group Domain Specific Review Board and SingHealth Centralized Institutional Review Board, Singapore (DSRB:D/2009/00021, CIRB:2018/2767/D).

Informed Consent Statement: Written informed consent was obtained from all subjects involved in this study. The children filled out an assent form to document their understanding of and participation in the study, while their parents gave written informed consent.

Data Availability Statement: Restrictions apply to the availability of some, or all, data generated or analyzed during this study. The corresponding author will, upon request, detail the restrictions and any conditions under which access to some data may be provided.

Acknowledgments: We thank the contributions of study participants, clinical and home-visit staff, and the GUSTO Study Group.

Conflicts of Interest: D.T.G. declares participation in the Scientific Advisory Board of Medela AG. Z.G., J.L.M., S.L.P and D.T.G. are/were supported by an unrestricted research grant from Medela AG, administered by The University of Western Australia. The funders had no role in the design of the study; in the collection, analyses, or interpretation of data; in the writing of the manuscript; or in the decision to publish the results. All other authors declare no conflicts of interest.

Reference

1. Katzman, D.K.; Bachrach, L.K.; Carter, D.R.; Marcus, R. Clinical and anthropometric correlates of bone mineral acquisition in healthy adolescent girls. *J. Clin. Endocrinol. Metab.* **1991**, *73*, 1332–1339. [CrossRef] [PubMed]

proceedings

MDPI

Abstract

Maternal Factors and Breast Anatomy and Milk Production during Established Lactation †

Zoya Gridneva [1,2,3,*], Alethea Rea [4], David Weight [4], Jacki L. McEachran [1,2,3], Ching Tat Lai [1,2,3], Sharon L. Perrella [1,2,3] and Donna T. Geddes [1,2,3]

1 School of Molecular Sciences, The University of Western Australia, Crawley, WA 6009, Australia; jacki.mceachran@uwa.edu.au (J.L.M.); ching-tat.lai@uwa.edu.au (C.T.L.); sharon.perrella@uwa.edu.au (S.L.P.); donna.geddes@uwa.edu.au (D.T.G.)
2 ABREAST Network, Perth, WA 6000, Australia
3 UWA Centre for Human Lactation Research and Translation, Crawley, WA 6009, Australia
4 Mathematics and Statistics, Murdoch University, Murdoch, WA 6150, Australia; alethea.rea@murdoch.edu.au (A.R.); david.weight@murdoch.edu.au (D.W.)
* Correspondence: zoya.gridneva@uwa.edu.au; Tel.: +61-8-6488-4467

† Presented at the Australian Breastfeeding + Lactation Research and Science Translation Conference (ABREAST Conference 2023), Perth, Australia, 10 November 2023.

Abstract: Animal models show a more rapid mammary gland response and more milk with subsequent lactations, as well as impairment of lactation performance by obesity. Whilst maternal obesity is linked to reduced breastfeeding initiation, breastfeeding confidence, and duration as well as early introduction of formula, maternal adiposity, breast anatomy and milk production (MP) have not been assessed in this population. Thirty-four lactating mothers 1–6 months postpartum and with BMI range of 17–35 kg/m^2 participated in this study. We conducted ultrasound examination imaging to assess breast anatomy. The amount of glandular tissue (glandular tissue representation (GTR)) was classified as low, moderate, or high. Number and diameters of milk ducts as well as mammary blood flow (the resistive index) were measured. Maternal bra cup volume was calculated from current bra size. Maternal body composition was measured with bioimpedance spectroscopy. Mothers completed a questionnaire regarding their medical, obstetric and lactation history, and conducted a 24 h MP study to enable calculation of total volume, average and maximum feed volumes and breast storage capacity (24 h MP parameters). For statistical analysis, we used the correlation networks method (directions of multiple significant correlations are reported). Correlation networks show that pathways culminating in either high or low MP start as early as puberty. In this study, later menarche correlates with the absence of breast growth during both puberty and pregnancy, which further correlate with lower numbers of ducts and smaller diameters. Higher maternal adiposity correlates with larger bra cup volume (both correlate with absence of breast growth during pregnancy and low GTR) and lower 24 h MP parameters. Larger numbers of ducts and duct diameters correlate with higher parity and longer durations of previous lactations, and higher 24 h MP parameters. Mammary blood flow shows no correlations. Findings from this cross-sectional study corroborate animal studies showing that a number of modifiable and non-modifiable maternal factors may impact breast development and MP. Further research may inform interventions, such as maintaining healthy adiposity not only during pre-conception, pregnancy, and lactation, but as early as childhood and potentially infancy. Moreover, the results provide rationale for antenatal lactation assessment of women and intervention in high-risk mothers to ensure they reach their full lactation potential.

Keywords: breastfeeding; lactation; milk production; breast anatomy; body composition; glandular tissue; milk ducts; ultrasound

Citation: Gridneva, Z.; Rea, A.; Weight, D.; McEachran, J.L.; Lai, C.T.; Perrella, S.L.; Geddes, D.T. Maternal Factors and Breast Anatomy and Milk Production during Established Lactation. *Proceedings* **2023**, *93*, 11. https://doi.org/10.3390/proceedings2023093011

Academic Editors: Debra J. Palmer and Nicolas L. Taylor

Published: 21 December 2023

Author Contributions: Conceptualization, Z.G. and D.T.G.; methodology, D.T.G., C.T.L., A.R., D.W. and Z.G.; data collection, D.T.G. and Z.G.; formal analysis, Z.G., A.R. and D.W.; investigation, D.T.G. and Z.G.; resources, D.T.G.; data curation, D.T.G., J.L.M. and Z.G.; writing—original draft preparation, Z.G.; writing—review and editing, A.R., D.W., J.L.M., C.T.L., S.L.P. and D.T.G.; visualization, Z.G.; supervision, D.T.G.; project administration, J.L.M.; funding acquisition, D.T.G. All authors have read and agreed to the published version of the manuscript.

Funding: This research was funded by unrestricted research grant from Medela AG (Switzerland). The funder had no role in the design of the study; in the collection, analyses, or interpretation of data; in the writing of the manuscript; or in the decision to publish the results.

Institutional Review Board Statement: The study was conducted in accordance with the Declaration of Helsinki. The study was approved by the Human Research Ethics Committee at The University of Western Australia (RA/4/20/6407) and conducted in accordance with the relevant guidelines and regulations.

Informed Consent Statement: Informed consent was obtained from all subjects involved in the study.

Data Availability Statement: Restrictions apply to the availability of some or all data generated or analyzed during this study. The corresponding author will on request detail the restrictions and any conditions under which access to some data may be provided.

Acknowledgments: We thank all our participants and their families for their time and help with this research.

Conflicts of Interest: D.T.G. declares participation in the Scientific Advisory Board of Medela AG. Z.G., J.L.M., C.T.L., S.L.P. and D.T.G. are/were supported by an unrestricted research grant from Medela AG, administered by The University of Western Australia. The funders had no role in the design of the study; in the collection, analyses, or interpretation of data; in the writing of the manuscript; or in the decision to publish the results. All other authors declare no conflict of interest.

proceedings

Abstract

A Comparison of the Efficacy and Comfort of Traditional and a Hands-Free, In-Bra Breastmilk Collection Pump Sets †

Zoya Gridneva *, Ashleigh H. Warden, Jacki L. McEachran, Ching Tat Lai, Sharon L. Perrella and Donna T. Geddes

School of Molecular Sciences, The University of Western Australia, Crawley, WA 6009, Australia; ashleigh.warden@uwa.edu.au (A.H.W.); jacki.mceachran@uwa.edu.au (J.L.M.); ching-tat.lai@uwa.edu.au (C.T.L.); sharon.perrella@uwa.edu.au (S.L.P.); donna.geddes@uwa.edu.au (D.T.G.)

* Correspondence: zoya.gridneva@uwa.edu.au; Tel.: +61-8-6488-4467

† Presented at the Australian Breastfeeding + Lactation Research and Science Translation Conference (ABREAST Conference 2023), Perth, Australia, 10 November 2023.

Keywords: lactation; human milk; breast pumping; breast expression; electric pump; comfortable pumping; hands-free pump; wearable pump

Breastfeeding women who pump their milk report that pumping is time-consuming [1] and interferes with mothering and other activities [2], and so there is high interest in effective wearable pumps. A recent study reported on the efficacy and comfort of an experimental hands-free, in-bra, breastmilk collection pump set (IBCPS) connected to a personal-use double electric breast pump (Freestyle), now available as the Freestyle Hands-Free Breast Pump [3]. Applying the IBCPS resulted in efficient and effective breast emptying with good comfort ratings. However, the performances of these two pump sets regarding efficacy, comfort, and pumping experience were not compared. This study focused on the data from a subset of participants that also completed the reference (Freestyle with a traditional shield) pumping session.

Twenty-one lactating mothers 1–6 months postpartum completed two pumping sessions with Freestyle, one using a traditional shield pump set and one with IBCPS. During the sessions, both breasts were pumped simultaneously (n = 42 sessions) using the participant's maximum comfortable vacuum for a period of 15 min of expression after the first milk ejection. Milk output was measured together with time to milk ejection and maternal comfort. A scale from 1 to 5 was used to rate participants' degrees of comfort and pumping experience perception in comparison with mothers' home pumps. A rating of 1 indicated (a) very comfortable and (b) much more comfortable, much more effective, much more likable, and much better compared with their home pump. A rating of 5 indicated (a) very uncomfortable and (b) much more uncomfortable, much more ineffective, much more unlikable, and much worse compared with their home pump. Mothers completed a 24 h milk production profile to allow the percentage of available milk removed to be calculated. Statistical analysis used linear mixed modelling accounting for degree of fullness of the breast pre-pumping, session, and a random effect of mother.

There was no significant difference between the two test conditions for volume of milk removed (IBCPS: 69 ± 39 g; Traditional: 69 ± 39 g), degree of fullness of the breast post-expression (IBCPS: 0.08 ± 0.13; Traditional: 0.08 ± 0.13), and percentage of available milk removed (IBCPS: 80 ± 57%; Traditional: 73 ± 28%). No significant difference in milk removal rate was seen (IBCPS: 4.7 ± 2.5 g/min; Traditional: 4.6 ± 2.7 g/min). However, time to milk ejection (by observation) was significantly shorter with IBCPS (IBCPS: 0.89 ± 0.28 min; Traditional: 1.25 ± 0.68 min, p = 0.002). Further, no significant differences were seen in nipple temperature changes (from pre- to post-pumping, IBCPS: −0.5 ± 1.1 °C; Traditional: −1.0 ± 1.1 °C), and initial comfort (IBCPS: 1.9 ± 0.6; Traditional: 2.1 ± 0.9),

Citation: Gridneva, Z.; Warden, A.H.; McEachran, J.L.; Lai, C.T.; Perrella, S.L.; Geddes, D.T. A Comparison of the Efficacy and Comfort of Traditional and a Hands-Free, In-Bra Breastmilk Collection Pump Sets. *Proceedings* **2023**, *93*, 12. https://doi.org/10.3390/proceedings2023093012

Academic Editors: Debra J. Palmer and Nicolas L. Taylor

Published: 21 December 2023

mid-pumping comfort (IBCPS: 1.9 ± 0.7; Traditional: 1.8 ± 0.7), and final comfort (IBCPS: 1.9 ± 0.9; Traditional: 1.8 ± 0.7) during sessions. Mothers had positive comments for IBCPS in general and compared with their home pumps. Most mothers found Freestyle with IBCPS to be more comfortable compared with their home pumps (IBCPS: 2.1 ± 0.9; Traditional: 2.8 ± 0.8, $p < 0.001$), though they did not rate it as effective as Freestyle with traditional shields (IBCPS: 2.3 ± 1.1; Traditional: 2.1 ± 0.9, $p < 0.001$). Mothers liked the in-bra, hands-free aspect better than the traditional (IBCPS: 1.4 ± 0.7; Traditional: 2.4 ± 0.9, $p < 0.001$) but did not think IBCPS was generally better than the Traditional pump set (IBCPS: 2.2 ± 1.0; Traditional: 2.6 ± 0.8) when compared to their home pumps.

Use of the IBCPS resulted in efficient and effective breast emptying without compromise in comfort. Mothers rated the IBCPS experience as comfortable and liked the in-bra, hands-free aspect and associated mobility. These findings indicate that use of effective and comfortable wearable pumps may assist women in supporting lactation while meeting their personal and career goals.

Author Contributions: Conceptualization, Z.G. and D.T.G.; methodology, D.T.G., C.T.L. and Z.G.; data collection, A.H.W., D.T.G. and Z.G.; formal analysis, Z.G. and A.H.W.; investigation, A.H.W., D.T.G. and Z.G.; resources, D.T.G.; data curation, A.H.W., D.T.G., J.L.M. and Z.G.; writing—original draft preparation, Z.G.; writing—review and editing, A.H.W., J.L.M., C.T.L., S.L.P. and D.T.G.; visualization, Z.G.; supervision, D.T.G.; project administration, J.L.M.; funding acquisition, D.T.G. All authors have read and agreed to the published version of the manuscript.

Funding: This research was funded by an unrestricted research grant from Medela AG (Switzerland). The funder had no role in the design of the study; in the collection, analyses, or interpretation of data; in the writing of the manuscript; or in the decision to publish the results.

Institutional Review Board Statement: The study was conducted in accordance with the Declaration of Helsinki. The study was approved by the Human Research Ethics Committee at The University of Western Australia (RA/4/20/6407) and conducted in accordance with the relevant guidelines and regulations.

Informed Consent Statement: Informed consent was obtained from all subjects involved in the study.

Data Availability Statement: Restrictions apply to the availability of some or all data generated or analyzed during this study. The corresponding author will on request detail the restrictions and any conditions under which access to some data may be provided.

Acknowledgments: We thank all our participants and their families for their time and help with this research.

Conflicts of Interest: D.T.G. declares participation in the Scientific Advisory Board of Medela AG. Z.G., J.L.M., C.T.L., S.L.P., A.H.W. and D.T.G. are/were supported by an unrestricted research grant from Medela AG, administered by The University of Western Australia. The funders had no role in the design of the study; in the collection, analyses, or interpretation of data; in the writing of the manuscript; or in the decision to publish the results.

References

1. Felice, J.P.; Geraghty, S.R.; Quaglieri, C.W.; Yamada, R.; Wong, A.J.; Rasmussen, K.M. "Breastfeeding" without baby: A longitudinal, qualitative investigation of how mothers perceive, feel about, and practice human milk expression. *Matern. Child Nutr.* **2017**, *13*, e12426. [CrossRef] [PubMed]
2. Flaherman, V.J.; Hicks, K.G.; Huynh, J.; Cabana, M.D.; Lee, K.A. Positive and negative experiences of breast pumping during the first 6 months. *Matern. Child Nutr.* **2016**, *12*, 291–298. [CrossRef] [PubMed]
3. Gridneva, Z.; Warden, A.H.; McEachran, J.L.; Lai, C.T.; Perrella, S.L.; Geddes, D.T. Human milk expression technologies: An evaluation of efficacy and comfort of hands-free, in-bra, breastmilk collection pump set. *Clin. Nutr. Open Sci.* **2023**, *49*, 28–36. [CrossRef]

proceedings

MDPI

Abstract

Maternal Perceptions of Sick/Preterm Infant Sleep and Settling Patterns in the First 9 Months †

Emma Shu Min Lim [1], Julie Williams [2], Philip Vlaskovsky [3], Demelza J. Ireland [1], Donna Tracy Geddes [4,5,6] and Sharon Lisa Perrella [4,5,6,*]

1 School of Biomedical Sciences, The University of Western Australia, Crawley, WA 6009, Australia; 22748392@student.uwa.edu.au (E.S.M.L.); demelza.ireland@uwa.edu.au (D.J.I.)
2 Neonatology Clinical Care Unit, King Edward Memorial Hospital, Subiaco, WA 6008, Australia; juliewill27a@gmail.com
3 Department of Mathematics and Statistics, School of Physics, Mathematics and Computing, The University of Western Australia, Crawley, WA 6009, Australia; philip.vlaskovsky@uwa.edu.au
4 School of Molecular Sciences, The University of Western Australia, Crawley, WA 6009, Australia; donna.geddes@uwa.edu
5 ABREAST Network, Perth, WA 6000, Australia
6 UWA Centre for Human Lactation Research and Translation, Crawley, WA 6009, Australia
* Correspondence: sharon.perrella@uwa.edu.au; Tel.: +61-64881208
† Presented at the Australian Breastfeeding + Lactation Research and Science Translation Conference (ABREAST Conference 2023), Perth, Australia, 10 November 2023.

Abstract: Infants born sick or preterm and admitted to the neonatal intensive care unit (NICU) face challenges to their sleep and the establishment of breastfeeding. The reported sleep patterns of NICU graduates are conflicting, and there is evidence of bi-directional relationships between infant feeding and sleep challenges and poor maternal mental health. A prospective observational longitudinal cohort study was conducted on mothers of sick/preterm infants with ≥ 5 days admission to the NICU. Participants were recruited at discharge, with follow up at 2 and 6 weeks, and 3, 6, and 9 months. Infant sleep, settle and cry patterns, maternal bother with infant behaviours, and maternal confidence were measured using the Sleep and Settle Questionnaire (SSQ). Feeding information was obtained, and breastfeeding confidence was measured using the Breastfeeding Self-Efficacy Scale–Short Form (BSES-SF). Our sample consisted of 94 mothers of 101 sick/preterm infants. The range of infant birth gestations was 23–42/40 weeks, with n = 37 (36.6%) born < 33/40 weeks, n = 26 (25.7%) born at 33 to 36^{+6}/40 weeks, and n = 38 (37.6%) born $\geq$ 37 weeks. The NICU length of stay was 5–173 days. Linear mixed modelling was used to examine associations between maternal bother and infant feeding and sleep variables. Night waking frequency ($p < 0.001$), durations of crying (day, evening, night; all $p \leq 0.001$), and durations of settling (evening, $p = 0.010$) were positively associated with increased maternal bother scores. Maternal confidence was negatively associated with maternal bother ($p < 0.001$). The median BSES-SF score at discharge was 59/70, with 49.5% of mothers no longer breastfeeding by 6 months. Maternal bother did not differ according to feeding method ($p = 0.44$), or birth gestation subgroup (i.e., born < 33 weeks, 33 to 36^{+6} weeks, or term sick) ($p = 0.44$). Whether NICU infants were term sick or preterm was not associated with maternal concern with infant sleep and settling patterns in the 9 months after hospital discharge. Mothers of sick/preterm infants reported greater bother with increased night waking and longer durations of crying and evening settling. The findings from this study can inform individualised anticipatory guidance and support for this vulnerable population.

Keywords: infant sleep; sleep behaviours; preterm infants; neonatal intensive care; breastfeeding; self-efficacy

Citation: Lim, E.S.M.; Williams, J.; Vlaskovsky, P.; Ireland, D.J.; Geddes, D.T.; Perrella, S.L. Maternal Perceptions of Sick/Preterm Infant Sleep and Settling Patterns in the First 9 Months. *Proceedings* **2023**, *93*, 13. https://doi.org/10.3390/proceedings2023093013

Academic Editors: Debra J. Palmer and Nicolas L. Taylor

Published: 22 December 2023

Author Contributions: Conceptualisation, S.L.P. and J.W.; methodology, S.L.P.; formal analysis, E.S.M.L. and P.V.; investigation, S.L.P. and E.S.M.L.; resources, D.T.G.; data curation, S.L.P.; writing—original draft preparation, E.S.M.L.; writing—review and editing, S.L.P., D.J.I. and D.T.G.; visualisation, P.V.; supervision, S.L.P., D.J.I. and D.T.G.; project administration, S.L.P.; funding acquisition, S.L.P. and D.T.G. All authors have read and agreed to the published version of the manuscript.

Funding: This research was funded by the Women and Infants' Research Foundation and Western Australian Nurses Memorial Charitable Trust. The salaries of D.T.G. and S.L.P. were funded by an unrestricted research grant from Medela AG (Switzerland) and administered by The University of Western Australia. The funders had no role in the design of the study; in the collection, analyses, or interpretation of data; in the writing of the manuscript; or in the decision to publish the results.

Institutional Review Board Statement: The study was conducted in accordance with the Declaration of Helsinki, and approved by the Ethics Committee of the Women's & Children's Health Service (reference EC06-05, October 2006).

Informed Consent Statement: Informed consent was obtained from all subjects involved in the study.

Data Availability Statement: The data that support the findings of this study are available from the corresponding author upon reasonable request.

Acknowledgments: We thank all of our participants for their time and help with this research.

Conflicts of Interest: D.T.G. declares participation in the Scientific Advisory Board of Medela AG. P.V., D.T.G. and S.L.P. are supported by an unrestricted research grant from Medela AG, administered by The University of Western Australia. The funders had no role in the design of the study; in the collection, analyses, or interpretation of data; in the writing of the manuscript; or in the decision to publish the results. All other authors declare no conflicts of interest.

proceedings

Abstract

Maternal Breast Growth and Body Mass Index in Relation to Milk Production †

Xuehua Jin [1,2,3], Ching Tat Lai [1,2,3], Sharon L. Perrella [1,2,3], Jacki L. McEachran [1,2,3], Zoya Gridneva [1,2,3] and Donna T. Geddes [1,2,3,*]

1 School of Molecular Sciences, The University of Western Australia, Crawley, WA 6009, Australia; xuehua.jin@research.uwa.edu.au (X.J.); ching-tat.lai@uwa.edu.au (C.T.L.); sharon.perrella@uwa.edu.au (S.L.P.); jacki.mceachran@uwa.edu.au (J.L.M.); zoya.gridneva@uwa.edu.au (Z.G.)
2 ABREAST Network, Perth, WA 6000, Australia
3 UWA Centre for Human Lactation Research and Translation, Crawley, WA 6009, Australia
* Correspondence: donna.geddes@uwa.edu.au
† Presented at the Australian Breastfeeding + Lactation Research and Science Translation Conference (ABREAST Conference 2023), Perth, Australia, 10 November 2023.

Keywords: lactation; human milk; milk production; body mass index; breast volume; breast growth

Low milk production poses a substantial challenge to exclusive and continued breastfeeding, which affects 10–15% of lactating mothers. This study aimed to explore the relationships between MP and both the maternal body mass index (BMI) and changes in breast volume during pregnancy and lactation.

Lactating mothers at 1–6 months postpartum conducted 24 h MP measurements using the test weighing method (n = 316; 74—low milk production (LMP), <600 mL; 242—normal milk production (NMP), ≥600 mL). Demographic and obstetric data, including maternal age, parity, infant sex, and birth mode, as well as pre-pregnancy and postpartum body mass index (BMI) and bra size, were provided. Maternal breast volume (cm^3) was calculated based on both bra cup size and band size. Descriptive statistics, Student's t-test, two-proportion z-test, correlation analysis and multivariable linear regression models were applied to elucidate maternal factors related to milk production.

The 24 h milk production in the LMP group was 466 ± 120 mL (80–599 mL) and 850 ± 191 mL (601–1682 mL) in the NMP group (p = 0.001). No significant differences were found between the groups in pre-pregnancy BMI, postpartum BMI, and BMI change (ΔBMI), as well as pre-pregnancy and postpartum breast volume, and breast volume increase (Δ breast volume). Both pre-pregnancy and postpartum BMI were positively correlated with pre-pregnancy and postpartum breast volume (r = 0.59–0.67, p = 0.001, for all). There was a higher proportion of mothers with Δ breast volume < 200 cm^3 in the LMP group compared with the NMP group (63% (41/65) vs. 45% (91/204)), respectively, p = 0.020). There was no difference in milk production between the LMP groups with Δ breast volume < 200 cm^3 and Δ breast volume > 200 cm^3. However, mothers in the LMP group with Δ breast volume < 200 cm^3 had lower milk production compared to both the NMP group with Δ breast volume < 200 cm^3 (mean difference ± standard error: 379 ± 34 mL, $p < 0.001$) and NMP group with Δ breast volume > 200 cm^3 (414 ± 33 mL, $p < 0.001$). Similarly, the LMP group with Δ breast volume > 200 cm^3 had a lower MP compared to both the NMP group with Δ breast volume < 200 cm^3 (343 ± 42 mL, $p < 0.001$) and NMP group with Δ breast volume > 200 cm^3 (377 ± 41 mL, $p < 0.001$). In addition, mothers in the LMP group with Δ breast volume > 200 cm^3 had a higher postpartum BMI compared to both the NMP group with Δ breast volume < 200 cm^3 (mean difference 4.1 ± 1.5, $p < 0.033$) and NMP group with Δ breast volume > 200 cm^3 (mean difference 4.6 ± 1.5, $p < 0.011$).

Citation: Jin, X.; Lai, C.T.; Perrella, S.L.; McEachran, J.L.; Gridneva, Z.; Geddes, D.T. Maternal Breast Growth and Body Mass Index in Relation to Milk Production. *Proceedings* **2023**, 93, 14. https://doi.org/10.3390/proceedings2023093014

Published: 22 December 2023

These preliminary results highlight the complex relationships between maternal BMI, increase in breast volume and 24 h milk production, and suggest that BMI and breast growth are potentially important indicators of milk production. Further investigations of these inter-related factors may inform interventions aiming at achieving normal milk production in women with higher adiposity.

Author Contributions: Conceptualization, X.J., C.T.L. and D.T.G.; methodology, C.T.L., J.L.M. and Z.G.; software, X.J. and C.T.L.; validation, X.J. and C.T.L.; formal analysis, X.J.; investigation, X.J. and C.T.L.; resources, S.L.P. and D.T.G.; data curation, X.J., C.T.L., J.L.M. and Z.G.; writing—original draft preparation, X.J.; writing—review and editing, C.T.L., S.L.P., Z.G. and D.T.G.; visualization, X.J.; supervision, C.T.L., S.L.P. and D.T.G.; project administration, D.T.G.; funding acquisition, D.T.G. All authors have read and agreed to the published version of the manuscript.

Funding: This research was funded by an unrestricted research grant from Medela AG (Switzerland), by The University of Western Australia (UWA) and China Scholarship Council (CSC). C.T.L., S.L.P., J.L.M., Z.G. and D.T.G. receive salaries from an unrestricted research grant paid by Medela AG and administered by UWA. X.J. is supported by UWA—CSC Joint PhD Scholarship and UWA—CSC Higher Degree by Research Top-Up Scholarship. The funders had no role in the design of the study; in the collection, analyses, or interpretation of data; in the writing of the manuscript; or in the decision to publish the results.

Institutional Review Board Statement: The study was conducted in accordance with the Declaration of Helsinki and was approved by the UWA Human Research Ethics Committee (2019/RA/4/20/4023).

Informed Consent Statement: Informed consent was obtained from all participants involved in the study.

Data Availability Statement: Data sharing is not applicable.

Acknowledgments: We are grateful to the families who participated in the study.

Conflicts of Interest: D.T.G. declares participation in the Scientific Advisory Board of Medela AG. C.T.L., S.L.P., J.L.M., Z.G., and D.T.G. are/were supported by an unrestricted research grant from Medela AG, administered by The University of Western Australia. The funders had no role in the design of the study; in the collection, analyses, or interpretation of data; in the writing of the manuscript; or in the decision to publish the results.

proceedings

MDPI

Abstract

Sampling Procedures for Estimating the Infant Intake of Human Milk Hormones, Glucose and Total Lipids †

Majed A. Suwaydi [1,2], Ching Tat Lai [1,3,4], Zoya Gridneva [1,3,4], Sharon L. Perrella [1,3,4], Mary E. Wlodek [5] and Donna T. Geddes [1,3,4,*]

1 School of Molecular Sciences, The University of Western Australia, Crawley, WA 6009, Australia; majed.suwaydi@reseach.uwa.edu.au or msuwaydi@jazanu.edu.sa (M.A.S.); ching-tat.lai@uwa.edu.au (C.T.L.); zoya.gridneva@uwa.edu.au (Z.G.); sharon.perrella@uwa.edu.au (S.L.P.)
2 Department of Medical Laboratory Technology, College of Applied Medical Sciences, Jazan University, Jazan 45142, Saudi Arabia
3 ABREAST Network, Perth, WA 6000, Australia
4 UWA Centre for Human Lactation Research and Translation, Crawley, WA 6009, Australia
5 Department of Obstetrics and Gynaecology, The University of Melbourne, Melbourne, VIC 3010, Australia; m.wlodek@unimelb.edu.au
* Correspondence: donna.geddes@uwa.edu.au; Tel.: +61-8-6488-4467

† Presented at the Australian Breastfeeding + Lactation Research and Science Translation Conference (ABREAST Conference 2023), Perth, Australia, 10 November 2023.

Abstract: Human milk (HM) components are highly variable, and infants consume different volumes of milk. There has been little focus on evaluating the effectiveness of protocols in estimating infant intake of milk components before studying their impact on infant outcomes. Our goal was to compare 24 h measured intake with estimated intakes from different sampling protocols in order to determine the most accurate method for estimating infant intakes of milk leptin, adiponectin, insulin, glucose and total lipids. Mothers of term infants (n = 20) collected pre- and post-feed samples and measured their infant milk intake during each feed over a 24 h period using the test weighing method at 3–6 months postpartum. Infant true intakes of HM leptin, adiponectin, insulin, glucose and total lipids were calculated by averaging the measured pre- and post-feed concentrations and multiplying by the milk intake for the corresponding feed. Intakes were then summed to provide total intake of each component over 24 h. The estimated intakes were calculated with concentrations determined using five different sampling protocols, designed to be representative of sampling protocols used in previous HM component studies: (a) morning pre-feed sampling, (b) morning post-feed sampling, (c) average of morning pre- and post-feed sampling, (d) average of three pre-feed samples from the morning (06:00–09:00), afternoon (13:00–16:00) and evening (19:00–22:00) and (e) average of six pre- and post-feed samples from the morning (06:00–09:00), afternoon (13:00–16:00) and evening (19:00–22:00). The concentration from each protocol was further multiplied by true measured intake, a constant average intake of 800 mL/24 h and a global average milk intake of 766 mL/24 h to obtain the estimated intakes (15 protocols). The average intake of HM was 791 ± 212 mL. Comparison revealed that using the average measured concentration from three sets of pre- and post-feed samples, taken in the morning (06:00–09:00), afternoon (13:00–16:00) and evening (19:00–22:00), multiplied by either true infant 24 h measured intake, a constant estimate of milk intake (800 mL) or global average of milk intake (766 mL) provided the most accurate estimation of the infant's intake for all components ($p > 0.05$). To obtain accurate estimates of HM leptin, adiponectin, insulin, glucose and total lipid intake in the absence of 24 h sampling, it is recommended to use a sampling protocol that involves taking samples before and after at least three breastfeeding sessions in a 24 h period.

Keywords: human milk; lactation; breastfeeding; leptin; adiponectin; insulin; glucose; fat; lipids; dose; infant nutrition; human milk intake

Citation: Suwaydi, M.A.; Lai, C.T.; Gridneva, Z.; Perrella, S.L.; Wlodek, M.E.; Geddes, D.T. Sampling Procedures for Estimating the Infant Intake of Human Milk Hormones, Glucose and Total Lipids. *Proceedings* **2023**, *93*, 15. https://doi.org/10.3390/proceedings2023093015

Academic Editors: Debra J. Palmer and Nicolas L. Taylor

Published: 22 December 2023

Author Contributions: Conceptualization, M.A.S., C.T.L. and D.T.G.; methodology, M.A.S., C.T.L. and D.T.G.; validation, M.A.S., C.T.L. and D.T.G.; formal analysis, M.A.S.; investigation, M.A.S., Z.G., C.T.L. and D.T.G.; resources, D.T.G.; data curation, Z.G. and S.L.P.; writing—original draft preparation, M.A.S.; writing—review and editing, C.T.L., Z.G., S.L.P., M.E.W. and D.T.G.; supervision, C.T.L., D.T.G. and M.E.W.; project administration, D.T.G.; funding acquisition, D.T.G. All authors have read and agreed to the published version of the manuscript.

Funding: This work was supported by Jazan University, Saudi Arabia, and an unrestricted research grant from Medela AG, Switzerland. M.A.S. receives a postgraduate scholarship from Jazan University. D.T.G., C.T.L., S.L.P. and Z.G. receive a salary from an unrestricted research grant from Medela AG that is administered by The University of Western Australia. The funders had no role in the design of the study; in the collection, analyses or interpretation of the data; in the writing of the manuscript; or in the decision to publish the results.

Institutional Review Board Statement: This study was approved by the Human Research of The University of Western Australia (UWA Human Research Ethics (RA/4/1/4492)).

Informed Consent Statement: Informed consent was obtained from all subjects involved in the study.

Data Availability Statement: Data sharing is not applicable.

Acknowledgments: We are grateful to the families who participated in the study.

Conflicts of Interest: D.T.G. declares participation in the Scientific Advisory Board of Medela AG. M.A.S. receives a postgraduate scholarship from Jazan University. D.T.G., C.T.L., Z.G. and S.L.P. receive a salary from an unrestricted research grant paid for by Medela AG and administered by The University of Western Australia. The funders had no role in the design of the study; in the collection, analyses or interpretation of data; in the writing of the manuscript; or in the decision to publish the results.

proceedings

MDPI

Abstract

Development of Breastfeeding Behaviours in Preterm Infants †

Melissa Klifunis [1], Demelza J. Ireland [1], Donna T. Geddes [2,3,4] and Sharon L. Perrella [2,3,4,*]

[1] School of Biomedical Sciences, The University of Western Australia, Crawley, WA 6009, Australia; 22714479@student.uwa.edu.au (M.K.); demelza.ireland@uwa.edu.au (D.J.I.)
[2] School of Molecular Sciences, The University of Western Australia, Crawley, WA 6009, Australia; don-na.geddes@uwa.edu.au
[3] ABREAST Network, Perth, WA 6000, Australia
[4] UWA Centre for Human Lactation Research and Translation, Crawley, WA 6009, Australia
* Correspondence: sharon.perrella@uwa.edu.au; Tel.: +61-64881208
† Presented at the Australian Breastfeeding + Lactation Research and Science Translation Conference (ABREAST Conference 2023), Perth, Australia, 10 November 2023.

Abstract: Breastfeeding is particularly important for vulnerable preterm infants as it provides protection from infections and reduces newborn mortality. However, preterm infants are often too immature to breastfeed after birth and may have medical conditions that require admission to the neonatal nursery. The published literature on the development of preterm feeding skills has focused mostly on bottle feeding. In order to better support breastfeeding after preterm birth, there is a need for evidence on the development of breastfeeding skills in preterm infants. The aim of this study was to examine breastfeeding skill development in a group of infants born at 25–33 weeks' gestation. Infants were assessed during weekly monitored breastfeeds from 33 weeks corrected gestational age (CGA) using the Preterm Infant Breastfeeding Behaviour Scale (PIBBS), and milk transfer was measured. Mothers rated PIBBS items—rooting, areolar grasp, latch to the breast, sucking, longest sucking burst and swallowing—and clinical staff performed test weights. Pearson correlation was used to assess changes in PIBBS scores items over time and associations between total PIBBS score and milk transfer volume. Total PIBBS scores at 33, 34 and 35 weeks' CGA were compared between groups of infants born at <30/40 and 30–33/40 weeks using Student's *t*-test. Our cohort consisted of 60 preterm mother–infant dyads recruited from the neonatal nurseries at King Edward Memorial Hospital between February 2015 and February 2016. A positive trend was found between increasing CGA and higher ratings for six PIBBS items: rooting ($R^2 = 0.08$, $F\ (1, 164) = 13.9$, $p < 0.001$), areolar grasp ($R^2 = 0.11$, $F\ (1, 164) = 21.0$, $p < 0.001$), latching ($R^2 = 0.14$, $F\ (1, 164) = 27.5$, $p < 0.001$), sucking ($R^2 = 0.14$, $F\ (1, 164) = 27.1$, $p < 0.001$), longest sucking burst ($R^2 = 0.17$, $F\ (1, 164) = 32.3$, $p < 0.001$) and swallowing ($R^2 = 0.14$, $F\ (1, 163) = 26.1$, $p < 0.001$). A higher total PIBBS score was associated with a higher milk transfer volume (mL) ($R^2 = 0.214$, $F\ (1, 164) = 44.8$, $p < 0.001$). When compared to infants born at 30–33 weeks' gestation, infants born at $25–29^{+6}$ weeks' gestation had similar PIBBS scores at 33 weeks' CGA (9.2 ± 3.6 vs. 9.5 ± 4.1, $p = 0.83$) and lower scores at 34 weeks' CGA (9.2 ± 3.4 vs. 11.7 ± 4.3, $p = 0.036$) and 35 weeks' CGA (12.3 ± 3.1 vs. 14.9 ± 3.5, $p = 0.031$). The development of preterm breastfeeding skills advances from 33 weeks CGA with wide inter-individual variation and slower progression observed in those born < 30 weeks' gestation. Therefore, an individualised approach to anticipatory guidance regarding breastfeeding progression during the neonatal nursery stay is needed. Findings from this study can contribute to the formation of breastfeeding information resources for clinical staff and parents of preterm infants.

Keywords: breastfeeding; preterm birth; neonatal intensive care

Citation: Klifunis, M.; Ireland, D.J.; Geddes, D.T.; Perrella, S.L. Development of Breastfeeding Behaviours in Preterm Infants. *Proceedings* **2023**, *93*, 16. https://doi.org/10.3390/proceedings2023093016

Academic Editors: Debra J. Palmer and Nicolas L. Taylor

Published: 22 December 2023

Author Contributions: Conceptualisation, S.L.P.; methodology, S.L.P.; formal analysis, M.K.; investigation, S.L.P. and M.K.; resources, D.T.G.; data curation, S.L.P.; writing—original draft preparation, M.K.; writing—review and editing, S.L.P. and D.J.I.; supervision, S.L.P., D.J.I. and D.T.G.; project administration, S.L.P.; funding acquisition, D.T.G. All authors have read and agreed to the published version of the manuscript.

Funding: This research was partially funded by an unrestricted research grant from CHANNEL 7 TELETHON TRUST. The salaries of D.T.G. and S.L.P. were funded by an unrestricted research grant from Medela AG (Switzerland) and administered by The University of Western Australia. The funders had no role in the design of the study; in the collection, analyses, or interpretation of data; in the writing of the manuscript; or in the decision to publish the results.

Institutional Review Board Statement: The study was conducted in accordance with the Declaration of Helsinki and approved by the Women and Newborn Health Services Ethics Committee (EC00350) at King Edward Memorial Hospital, PRN #RGS0000002655.

Informed Consent Statement: Informed consent was obtained from all subjects involved in the study.

Data Availability Statement: The data that support the findings of this study are available from the corresponding author upon reasonable request.

Acknowledgments: We thank all of our participants for their time and help with this research.

Conflicts of Interest: D.T.G. declares participation in the Scientific Advisory Board of Medela AG. D.T.G. and S.L.P. are supported by an unrestricted research grant from Medela AG, administered by The University of Western Australia. The funders had no role in the design of the study; in the collection, analyses, or interpretation of data; in the writing of the manuscript; or in the decision to publish the results. All other authors declare no conflict of interest.

proceedings

MDPI

Abstract

Risk Factors for Low Milk Production [†]

Donna T. Geddes [1,2,3,*], Stuart A. Prosser [1,2,3,4], Zoya Gridneva [1,2,3] and Sharon L. Perrella [1,2,3,4]

1 School of Molecular Sciences, The University of Western Australia, Crawley, WA 6009, Australia; stuart@westernobs.com.au (S.A.P.); zoya.gridneva@uwa.edu.au (Z.G.); sharon.perrella@uwa.edu.au (S.L.P.)
2 ABREAST Network, Perth, WA 6000, Australia
3 UWA Centre for Human Lactation Research and Translation, Crawley, WA 6009, Australia
4 Western Obstetrics, Balcatta, WA 6021, Australia
* Correspondence: donna.geddes@uwa.edu.au; Tel.: +61-8-6488-7006

† Presented at the Australian Breastfeeding + Lactation Research and Science Translation Conference (ABREAST Conference 2023), Perth, Australia, 10 November 2023.

Abstract: Numerous factors not consistently identified in pregnancy are linked with decreased breastfeeding exclusivity and durations. These factors may be considered in three domains: the anatomical, metabolic, and psychosocial domains. As fundamental research into lactation has increased, it is now often possible to identify or speculate the mechanisms by which these factors potentially reduce milk production. The first domain describes the anatomical characteristics of the breast, including intrinsic factors such as hypoplasia (underdevelopment), which may have a genetic component and can be masked by breast augmentation surgery. Hypoplasia has long been associated with the inability to make a full milk production that satisfyies the infant's needs, although it is not possible to predict a woman's 24-h milk production so that appropriate complementary feeds can be advised. Extrinsic causes such as breast reduction surgery impact the volume of glandular tissue, thereby reducing the synthetic capacity of the breast. Whereas nipple piercings may damage milk ducts, obstructing milk flow from the breast and thereby reducing milk supply via the autocrine pathway. Various maternal metabolic disorders (intrinsic) comprise the second domain, which includes conditions such as gestational diabetes mellitus, type 1 and 2 diabetes, polycystic ovarian syndrome (often undiagnosed), and hypothyroidism. The aberrant levels of hormones associated with these disorders, such as insulin, are also implicated in breast development, raising the possibility of reduced mammary growth in pregnancy and, consequently, milk production. Much more research is needed in this area, not only to understand mechanisms by which lactation is impacted but also to identify the women at risk of reduced lactation capacity. The third and final domain includes psychosocial issues such as short intended breastfeeding durations, a lack of breastfeeding support, and maternal anxiety and depression. With respect to anxiety and depression, their association with reduced breastfeeding is likely multifaceted, encompassing mood and the potential biochemical changes associated with these states, such as lower levels of circulating oxytocin and higher cortisol levels. Possessing a knowledge of the negative impacts of the intrinsic and extrinsic factors within the maternal anatomical, metabolic, and psychosocial domains provides the impetus for antenatal lactation screening. The antenatal identification of risk factors enables anticipatory guidance and education during pregnancy, as well as early postpartum intervention should breastfeeding issues occur.

Keywords: breastfeeding; lactation; low milk production; breast anatomy; glandular tissue; risk factors

Citation: Geddes, D.T.; Prosser, S.A.; Gridneva, Z.; Perrella, S.L. Risk Factors for Low Milk Production. *Proceedings* **2023**, *93*, 17. https://doi.org/10.3390/proceedings2023093017

Academic Editors: Debra J. Palmer and Nicolas L. Taylor

Published: 26 December 2023

Author Contributions: Conceptualization, D.T.G., S.A.P., Z.G. and S.L.P.; methodology, D.T.G.; investigation, D.T.G., Z.G. and S.L.P.; writing—original draft preparation, D.T.G., S.A.P., Z.G. and S.L.P.; writing—review and editing, D.T.G., S.A.P., Z.G. and S.L.P.; project administration, D.T.G.; funding acquisition, D.T.G., Z.G. and S.L.P. All authors have read and agreed to the published version of the manuscript.

Funding: This study was funded by an unrestricted research grant from Medela AG (Switzerland).

Institutional Review Board Statement: Not applicable.

Informed Consent Statement: Not applicable.

Data Availability Statement: Data sharing is not applicable.

Conflicts of Interest: D.T.G. declares that she receives a salary from a research grant from Medela AG, administered by The University of Western Australia, and declares participation in the Scientific Advisory Board of Medela AG. Z.G. and S.L.P. receive a salary from an unrestricted research grant from Medela AG, administered by The University of Western Australia. S.A.P. declares no conflicts of interest.

proceedings

Abstract

Macronutrient Content in Human Milk Is Not Affected by Infant's Sex †

Karina D. Honoré [1,2,*], Signe Bruun [3], Sören Möller [4], Kim F. Michaelsen [5], Steffen Husby [1,2] and Gitte Zachariassen [1,2]

1 Department of Clinical Research, Faculty of Health Science, University of Southern Denmark, 5000 Odense, Denmark; steffen.husby@rsyd.dk (S.H.); gitte.zachariassen@rsyd.dk (G.Z.)
2 Hans Christian Andersen Children's Hospital, Odense University Hospital, 5000 Odense, Denmark
3 Department of Anesthesiology, Odense University Hospital, 5000 Odense, Denmark; signe.bruun@rsyd.dk
4 Research Unit OPEN, Department of Clinical Research, University of Southern Denmark, 5000 Odense, Denmark; soren.moller@rsyd.dk
5 Department of Nutrition, Exercise and Sports, University of Copenhagen, 1958 Frederiksberg, Denmark; kfm@nexs.ku.dk
* Correspondence: karina.dyrvig.honore@rsyd.com; Tel.: +45-29-90-30-68
† Presented at the Australian Breastfeeding + Lactation Research and Science Translation Conference (ABREAST Conference 2023), Perth, Australia, 10 November 2023.

Abstract: Human milk contains macronutrients possibly affecting infant and early childhood growth. Most studies suggest a sex-specific difference in macronutrient levels in favor of males, with a higher energy content from fat, lactose, and protein. Further, macronutrient levels may differ according to mixed or exclusive breastfeeding; however, the literature is inconsistent. Our aim was to investigate sex-specific differences in macronutrient content in exclusively breastfed infants in a Danish child cohort, and whether macronutrient levels differed between exclusive or mixed (breast and formula) breastfeeding. Participants were part of the prospective birth cohort Odense Child Cohort. Baseline characteristics were obtained from medical records. Weekly SMS questions were sent to the mothers until the cessation of breastfeeding, asking whether they were breastfeeding and/or formula feeding. Mothers delivered a milk sample at the planned 3–4-month examination of the infant. Macronutrient analyses were performed on 182 samples using mid-infrared transmission spectroscopy (Miris Human milk Analyzer). We included 150 mother–infant dyads with both macronutrient analysis, and SMS data on breastfeeding. Baseline characteristics did not differ according to sex. The median interquartile range (IQR) infant age at the time of sampling was 4.1 (3.7–4.5) months. A total of 39 males and 38 females were exclusively breastfed at the time of milk sampling, while 36 males and 37 females were mixed-fed. We found no significant sex-specific differences in macronutrients among exclusively breastfed infants. The median (IQR) levels for males and females, respectively, were; protein, 0.85 g/100 mL (0.77, 0.90), and 0.82 g/100 mL (0.80, 0.90), $p = 0.91$; lactose, 7.83 g/100 mL (7.70, 7.95), and 7.73 g/100 mL (7.53, 7.90), $p = 0.17$; fat, 3.23 g/100 mL (2.07, 4.37), and 3.07 g/100 mL (2.10, 3.60), $p = 0.34$; energy, 65.5 kcal/100 mL (54.17, 77.00), and 63 kcal/100 mL (56.00, 69.33), $p = 0.13$. Further, we found no significant differences in macronutrient content in human milk samples from exclusively versus mixed-feeding mothers either prior and after adjusting for confounders, $p > 0.36$. This study does not confirm the previous findings of sex-specific differences in macronutrients in human milk. It is still unknown if sex-specific formula products tailored to meet possible sex-specific requirements can optimize child growth. Further research on this topic is needed.

Keywords: human milk; sex-specific differences; macronutrient content; exclusive breastfeeding; mixed breastfeeding

Citation: Honoré, K.D.; Bruun, S.; Möller, S.; Michaelsen, K.F.; Husby, S.; Zachariassen, G. Macronutrient Content in Human Milk Is Not Affected by Infant's Sex. *Proceedings* **2024**, *93*, 18. https://doi.org/10.3390/proceedings2023093018

Academic Editors: Zoya Gridneva and Donna T. Geddes

Published: 27 December 2023

Author Contributions: Conceptualization, K.D.H., S.B., and G.Z.; methodology, K.D.H., S.B., S.M., and G.Z.; Formal analysis, K.D.H.; investigation K.D.H.; resources, K.D.H.; data curation, K.D.H.,

S.B., and S.M.; writing—original draft preparation, K.D.H.; writing—review and editing, K.D.H., S.B., S.M, K.F.M., S.H., and G.Z.; supervision, G.Z.; project administration, K.D.H.; funding acquisition, K.D.H. All authors have read and agreed to the published version of the manuscript.

Funding: This research was partially funded by the Region of Southern Denmark, the Faculty Scholarship from the University of Southern Denmark, and Arla Foods Ingredients.

Institutional Review Board Statement: The study was approved by The Regional Scientific Ethical Committee for Southern Denmark (project number S-20090130, sub-protocols 12 and 18) and The Danish Data Protection Agency (case number 12/26892). The study complied with the Declaration of Helsinki.

Informed Consent Statement: Informed consent was obtained from all subjects involved in the study.

Data Availability Statement: The data presented in this study are available on request from the corresponding author.

Acknowledgments: Thanks to all participating children and their families, as well as the employees at the Odense Child Cohort. Special thanks to scientific laboratory technician Mette Vogn Hviid for helping with the collection and processing of the milk samples.

Conflicts of Interest: K.D.H. has received funding from Arla Foods Ingredients and has received speaker fees from Nestlé Nutrition Institute. G.Z. has received speaker fees from Nestlé Nutrition Institute and Mead Johnson Nutrition Institute. S.B. was employed by Arla Foods Ingredients from February 2016 to January 2019 and has received speaker fees from Ordesa Group and Semper. K.F.M. has received research grants from the US Dairy Export Council and the Danish Dairy Research Foundation and has undertaken research in collaboration with Nutriset. S.H. has received unrestricted grants from Thermo-Fisher and Takeda and has been on an advisory board for Glaxo-Smith-Kline. The funders had no role in the design of the study; in the collection, analyses, or interpretation of data; in the writing of the manuscript; or in the decision to publish the results.

proceedings

Abstract

Domperidone: Pharmacists Stimulating Clinical Change for Lactation Consultants †

Katherine Chinnery [1,*], Stephanie Wai Khuan Teoh [1], Tamara Lebedevs [1] and Myra Kildunne [2]

1 Pharmacy Department, King Edward Memorial Hospital, Subiaco 6008, Australia; stephanie.teoh@health.wa.gov.au (S.W.K.T.); tamara.lebedevs@health.wa.gov.au (T.L.)
2 Breastfeeding Centre, King Edward Memorial Hospital, Subiaco 6008, Australia; myra.kildunne@health.wa.gov.au
* Correspondence: katherine.chinnery@health.wa.gov.au; Tel.: +618-64582222
† Presented at the Australian Breastfeeding + Lactation Research and Science Translation Conference (ABREAST Conference 2023), Perth, Australia, 10 November 2023.

Abstract: Domperidone is a commonly prescribed galactagogue used off-label for lactation insufficiency. Prescriber unfamiliarity or safety concerns can lead to therapeutic delay and potential early breastfeeding discontinuation. To facilitate access, the study site pharmacy department developed a structured administration and supply arrangement (SASA) for International Board Certified Lactation Consultants to screen and initiate domperidone using a checklist. The study aimed to validate a domperidone screening tool via an analysis of its use and compliance. Records were extracted from the RedCAP® database for the first 50 women with a documented domperidone supply and reviewed against medical records. A staff survey was distributed assessing compliance and attitudes towards the SASA. Records of supply from the RedCAP® database revealed 34% (17/50) of patients were referred to a physician, revealing a discrepancy between RedCAP® reporting and checklists as no referrals were documented. Overall staff satisfaction with the SASA was rated 4.6/5. In total, 77.7% (7/9) felt confident counselling and supplying domperidone with the SASA in place, and 88.9% (8/9) felt confident using the checklist to identify the appropriateness of therapy and referral to a physician. Only 55.6% (5/9) indicated the checklist was used with each screening. The SASA education package is being updated to clarify the requirements for checklist completion and standardise frameworks to document follow-up.

Keywords: domperidone; galactagogue; lactation insufficiency; lactation consultant

Citation: Chinnery, K.; Teoh, S.W.K.; Lebedevs, T.; Kildunne, M. Domperidone: Pharmacists Stimulating Clinical Change for Lactation Consultants. *Proceedings* **2023**, 93, 19. https://doi.org/10.3390/proceedings2023093019

Academic Editors: Zoya Gridneva and Donna T. Geddes

Published: 3 January 2024

Author Contributions: Conceptualization, S.W.K.T. and T.L.; methodology, S.W.K.T. and T.L.; formal analysis, K.C.; data curation, K.C.; writing—original draft preparation, K.C.; writing—review and editing, S.W.K.T., T.L. and M.K. All authors have read and agreed to the published version of the manuscript.

Funding: This research received no external funding.

Institutional Review Board Statement: The study was conducted in accordance with the Declaration of Helsinki and approved by the Women and Newborn Health Service Quality Improvement Committee of King Edward Memorial Hospital on 7 March 2023 (approval number: GEKO 49245).

Informed Consent Statement: Participant Information Form was obtained from all staff members involved in the staff survey. Patient consent was waived with deidentified patient information in the study and deemed Low and/or Negligible Risk Research/Activities by the Quality Improvement Committee of King Edward Memorial Hospital.

Data Availability Statement: Data are contained within the article.

Acknowledgments: The authors would like to acknowledge the contribution of the pharmacy staff and International Board Certified Lactation Consultants at King Edward Memorial Hospital in documenting their valuable and critical work for the health service.

Conflicts of Interest: The authors declare no conflicts of interest.

proceedings

MDPI

Abstract

What Women Want: Supporting Breastfeeding after Caesarean Birth †

Sharon L. Perrella [1,2,3,*], Sarah G. Abelha [4], Jacki L. McEachran [1,2,3], Philip Vlaskovsky [5], Stuart A. Prosser [1,4] and Donna T. Geddes [1,2,3]

1 School of Molecular Sciences, The University of Western Australia, Crawley, WA 6009, Australia; jacki.mceachran@uwa.edu.au (J.L.M.); stuart@westernobs.com.au (S.A.P.); donna.geddes@uwa.edu.au (D.T.G.)
2 ABREAST Network, Perth, WA 6000, Australia
3 UWA Centre for Human Lactation Research and Translation, Crawley, WA 6009, Australia
4 Western Obstetrics, Balcatta, WA 6021, Australia; sarahabelha@live.com.au
5 Department of Mathematics and Statistics, School of Physics, Mathematics and Computing, The University of Western Australia, Crawley, WA 6009, Australia; philip.vlaskovsky@uwa.edu.au
* Correspondence: sharon.perrella@uwa.edu.au; Tel.: +61-8-6488-4467

† Presented at the Australian Breastfeeding + Lactation Research and Science Translation Conference (ABREAST Conference 2023), Perth, Australia, 10 November 2023.

Keywords: breastfeeding; caesarean section; postpartum; postcaesarean section; midwife

Caesarean birth is associated with suboptimal breastfeeding outcomes such as delayed initiation, a higher incidence of difficulties and shorter duration when compared to vaginal birth. Contributing factors may include post-operative pain and reduced mobility as well as higher rates of maternal and infant health complications. Breastfeeding experiences may also differ between women that have elective and non-elective caesarean births as well as between primiparous and multiparous women. Exploration of women's experiences of establishing breastfeeding after caesarean birth is needed to better understand their needs.

A mixed-methods descriptive study was conducted using an anonymous online questionnaire to determine Australian women's experiences of establishing breastfeeding in hospital and during the first 2 weeks at home after caesarean birth. Thematic analysis of qualitative data was used to identify themes relating to breastfeeding establishment. Inclusion criteria were birth at ≥37 weeks gestation within the previous 12 months, and age ≥ 18 years.

Questionnaires were completed by n = 961 women that were 33.0 ± 6.2 years of age, n = 480 (50%) primiparous, and 6.1 ± 4.3 months postpartum. Reported caesarean birth types were non-elective, 42.9%; elective for medical indications, 39.5%; and elective by maternal request, 17.6%. Initiation of breastfeeding occurred within an hour of birth for n = 675 women (70.2%), and n = 771 (80.2%) breastfed their infant during the postnatal ward stay. One third of women (n = 275, 32%) reported it was easy to pick up their baby to breastfeed and half (n = 418, 49%) received conflicting information from hospital staff. Qualitative themes relating to breastfeeding establishment included experiences of clinical care, maternal expectations and difficulty in picking up the baby. Aspects of care found to be helpful included midwives spending time with women to show and explain breastfeeding, consistent advice and recognition of multiparous women that needed breastfeeding support. Physical and emotional support of the partner was highly valued.

In the days following caesarean birth, women benefit from physical support in accessing the baby for breastfeeding, while availability and consistency of clinical lactation support is vital. As the current global midwifery shortage poses challenges to staffing, inclusion of the partner in breastfeeding education and extension of their 'visiting hours' may

Citation: Perrella, S.L.; Abelha, S.G.; McEachran, J.L.; Vlaskovsky, P.; Prosser, S.A.; Geddes, D.T. What Women Want: Supporting Breastfeeding after Caesarean Birth. *Proceedings* **2023**, *93*, 20. https://doi.org/10.3390/proceedings2023093020

Published: 9 January 2024

improve breastfeeding support, while re-examination of the education of all postpartum health care providers is needed to ensure consistent evidence-based lactation care.

Author Contributions: Conceptualization, S.L.P., S.A.P. and D.T.G.; methodology, S.L.P., S.G.A. and J.L.M.; data collection, J.L.M.; formal analysis, S.G.A., S.L.P. and P.V.; investigation, S.L.P. and S.G.A.; resources, D.T.G.; data curation, J.L.M.; writing—original draft preparation, S.L.P.; writing—review and editing, S.A.P. and D.T.G.; supervision, S.L.P. and D.T.G.; project administration, J.L.M.; funding acquisition, D.T.G. All authors have read and agreed to the published version of the manuscript.

Funding: This research was funded by unrestricted research grant from Medela AG (Switzerland).

Institutional Review Board Statement: The study was conducted in accordance with the Declaration of Helsinki. The study was approved by the Human Research Ethics Committee at The University of Western Australia (2022/ET000174) and conducted in accordance with the relevant guidelines and regulations.

Informed Consent Statement: Informed consent was obtained from all subjects involved in the study.

Data Availability Statement: Restrictions apply to the availability of some, or all data generated or analyzed during this study. The corresponding author will on request detail the restrictions and any conditions under which access to some data may be provided.

Acknowledgments: We thank all our participants for their time and help with this research.

Conflicts of Interest: D.T.G. declares participation in the Scientific Advisory Board of Medela AG. S.L.P. and D.T.G. are supported by an unrestricted research grant from Medela AG, administered by The University of Western Australia. The funders had no role in the design of the study; in the collection, analyses, or interpretation of data; in the writing of the manuscript; or in the decision to publish the results. All other authors declare no conflicts of interest.

proceedings

Abstract

Comparing Breastfeeding Outcomes of Australian Women Who Birth by Elective and Non-Elective Caesarean Section †

Lauren Papalia [1,2], Sarah G. Abelha [3], Jacki L. McEachran [4,5,6], Stuart J. Watson [1,2,7], Linda McKean [1,2], Stuart A. Prosser [3,4,5,6], Donna T. Geddes [4,5,6] and Sharon L. Perrella [4,5,6,*]

1 School of Nursing, Curtin University, Bentley, WA 6102, Australia; 12874417@student.curtin.edu.au (L.P.); stuart.watson@health.wa.gov.au (S.J.W.); linda.mckean@health.wa.gov.au (L.M.)
2 King Edward Memorial Hospital, Subiaco, WA 6008, Australia
3 Western Obstetrics, Balcatta, WA 6021, Australia; sarahabelha@live.com.au (S.G.A.); stuart@westernobs.com.au (S.A.P.)
4 School of Molecular Sciences, The University of Western Australia, Crawley, WA 6009, Australia; jacki.mceachran@uwa.edu.au (J.L.M.); donna.geddes@uwa.edu (D.T.G.)
5 ABREAST Network, Perth, WA 6000, Australia
6 UWA Centre for Human Lactation Research and Translation, Crawley, WA 6009, Australia
7 Psychology, Murdoch University, Murdoch, WA 6150, Australia
* Correspondence: sharon.perrella@uwa.edu.au; Tel.: +61-64881208
† Presented at the Australian Breastfeeding + Lactation Research and Science Translation Conference (ABREAST Conference 2023), Perth, Australia, 10 November 2023.

Abstract: Caesarean section birth (CS) accounts for 38% of Australian births and is known to negatively impact breastfeeding exclusivity and duration. Comparison of breastfeeding outcomes between elective (ELUSCS) and non-elective lower uterine segment caesarean section (NELUSCS) births is needed to inform clinical care. Secondary analysis was performed on data from Australian women who had birthed by CS within the previous 12 months and completed an anonymous online questionnaire. Women responded to items relating to CS birth type and feeding methods immediately post birth, during the hospital stay and in the first 2 weeks at home. Self-reported pain at those time points was rated using a scale of 0–10 with 0 indicating no pain and 10 indicating severe pain. Associations between CS birth type, pain scores and feeding methods were determined, and breastfeeding prevalence at the time of survey completion examined. Our sample consisted of 851 women at 5 ± 3.5 months postpartum, of which 435 (51.1%) were primiparous and 362 (42%) had a NELUSCS birth. Infants born by NELUSCS were more likely to receive formula (37.1% vs. 28.8% ELUSCS, $p = 0.024$) and less likely to be breastfed (78.6% vs. 85.9% ELUSCS, $p = 0.022$) during the hospital stay. During the 2 weeks after discharge, women who birthed by NELUSCS had higher mean pain scores (6.5 vs. 4.6 ELUSCS, $p < 0.001$). Few women expressed and fed their milk in the weeks after discharge, with higher rates seen after NELUSCS (7.7% vs. 3.5%, $p = 0.017$). Breastfeeding status at study completion was not associated with CS birth type (19.6% ELUSCS vs. 19% NELUSCS, $p = 0.99$). The study findings indicate that Australian women who give birth by NELUSCS are more likely to experience breastfeeding challenges, with higher rates of infant formula supplementation and a more painful recovery in the days and weeks after birth. In light of the higher postpartum pain scores and lower rates of exclusive breastfeeding, women who birth by NELUSCS need additional postpartum support.

Keywords: caesarean section; elective; non elective; breastfeeding; infant formula; pain

Citation: Papalia, L.; Abelha, S.G.; McEachran, J.L.; Watson, S.J.; McKean, L.; Prosser, S.A.; Geddes, D.T.; Perrella, S.L. Comparing Breastfeeding Outcomes of Australian Women Who Birth by Elective and Non-Elective Caesarean Section. *Proceedings* **2023**, *93*, 21. https://doi.org/10.3390/proceedings2023093021

Academic Editors: Debra J. Palmer and Nicolas L. Taylor

Published: 9 January 2024

Author Contributions: Conceptualization, S.L.P., S.G.A., S.A.P. and D.T.G.; methodology, S.L.P., S.G.A. and J.L.M.; software, J.L.M.; formal analysis, L.P. and S.J.W.; investigation, S.L.P. and S.G.A.; resources, D.T.G.; data curation, J.L.M.; writing—original draft preparation, L.P.; writing—review and editing, S.L.P., D.T.G. and L.M.; supervision, S.L.P. and L.M.; project administration, L.P. All authors have read and agreed to the published version of the manuscript.

Funding: This research received no external funding. The salaries of D.T.G. and S.L.P. were funded by an unrestricted research grant from Medela AG (Switzerland) and administered by The University of Western Australia. The funders had no role in the design of the study; in the collection, analyses, or interpretation of data; in the writing of the manuscript; or in the decision to publish the results.

Institutional Review Board Statement: The study was conducted in accordance with the Declaration of Helsinki and approved by the Institutional Review Board (or Ethics Committee) of Human Research Ethics Committee of The University of Western Australia (approval number: 2022/ET000174 10/08/2023). Reciprocal ethics approval was granted from Curtin University Human Research Ethics Office (approval number: HRE2023-0505, approval date 07/09/2023).

Informed Consent Statement: Informed consent was obtained from all subjects involved in the study.

Data Availability Statement: The data that support the findings of this study are available from the corresponding author upon reasonable request.

Acknowledgments: We thank all of our participants for their time and help with this research.

Conflicts of Interest: D.T.G. declares participation in the Scientific Advisory Board of Medela AG. D.T.G. and S.L.P. are supported by an unrestricted research grant from Medela AG, administered by The University of Western Australia. The funders had no role in the design of the study; in the collection, analyses, or interpretation of data; in the writing of the manuscript; or in the decision to publish the results. All other authors declare no conflicts of interest.

proceedings

Abstract

Maternal Dietary Intervention during Lactation Impacts the Maternal Faecal and Human Milk Microbiota †

Azhar S. Sindi [1,2,3,4], Lisa F. Stinson [3,4,5], Zoya Gridneva [3,4,5], Gabriela E. Leghi [6], Merryn J. Netting [7,8,9], Mary E. Wlodek [5,10], Beverly S. Muhlhausler [6,11], Alethea Rea [12,13], Michelle L. Trevenen [12], Donna T. Geddes [3,4,5] and Matthew S. Payne [1,*]

1 Division of Obstetrics and Gynaecology, School of Medicine, The University of Western Australia, Perth, WA 6008, Australia; azhar.sindi@research.uwa.edu.au or asmsindi@uqu.edu.sa
2 College of Applied Medical Sciences, Umm Al-Qura University, Makkah 24381-8156, Saudi Arabia
3 ABREAST Network, Perth, WA 6000, Australia; lisa.stinson@uwa.edu.au (L.F.S.); zoya.gridneva@uwa.edu.au (Z.G.); donna.geddes@uwa.edu.au (D.T.G.)
4 UWA Centre for Human Lactation Research and Translation, Perth, WA 6009, Australia
5 School of Molecular Sciences, The University of Western Australia, Perth, WA 6009, Australia; mary.wlodek@uwa.edu.au or m.wlodek@unimelb.edu.au
6 School of Agriculture, Food and Wine, The University of Adelaide, Adelaide, SA 5064, Australia; gabriela.leghivoyer@gmail.com (G.E.L.); beverly.muhlhausler@adelaide.edu.au or bev.muhlhausler@csiro.au (B.S.M.)
7 Women and Kids Theme, South Australian Health and Medical Research Institute (SAHMRI), Adelaide, SA 5000, Australia; merryn.netting@adelaide.edu.au
8 Discipline of Paediatrics, The University of Adelaide, Adelaide, SA 5006, Australia
9 Women's and Children's Hospital, Adelaide, SA 5006, Australia
10 Department of Obstetrics and Gynaecology, University of Melbourne, Melbourne, VIC 3010, Australia
11 CSIRO, Adelaide, SA 5000, Australia
12 Centre for Applied Statistics, The University of Western Australia, Perth, WA 6009, Australia; alethea.rea@murdoch.edu.au (A.R.); michelle.trevenen@uwa.edu.au (M.L.T.)
13 Mathematics and Statistics, Murdoch University, Murdoch, WA 6150, Australia
* Correspondence: matthew.payne@uwa.edu.au
† Presented at the Australian Breastfeeding + Lactation Research and Science Translation Conference (ABREAST Conference 2023), Perth, Australia, 10 November 2023.

Abstract: Diet is a key factor that shapes the gut microbiome. Maternal diet has been proposed as a potential modulator of the human milk microbiome. However, the effect of diet during lactation on the maternal microbiota remains unclear. This study, therefore, set out to determine the effect of a two-week reduced fat and sugar and increased fibre maternal dietary intervention on the maternal faecal and milk microbiota. Faecal swabs and human milk samples were collected from mothers ($n = 11$) immediately pre-intervention, immediately post-intervention, and 4- and 8-weeks post-intervention, and were analysed using full-length 16S rRNA gene sequencing. The maternal macronutrient intake was assessed across one week prior to the intervention using 24 h dietary recall and during the intervention using FoodWorks 10 Software. The maternal fat and sugar intake significantly decreased from pre-intervention (fat: 120.9 ± 39.4 g; sugar: 114.1 ± 40.9 g) to the first (fat: 52.9 ± 4.3 g, $p < 0.001$; sugar: 83.2 ± 5.1 g, $p = 0.005$) and second week of the intervention (fat: 52.3 ± 6.2 g, $p < 0.001$; sugar: 82.7 ± 6.5 g, $p = 0.005$). The dietary fibre intake significantly increased from pre-intervention (28.8 ± 8.3 g) to the first week of the intervention (34.6 ± 2.8 g, $p = 0.012$) but was not different in the second week of the intervention compared to pre-intervention. Significant changes in the bacterial composition of maternal faeces were detected after the dietary intervention, with decreases in the relative abundance of *Bacteroides caccae* and increases in the relative abundance of *Faecalibacillus intestinalis*. In human milk, a significant increase in *Cutibacterium acnes* and a decrease in *Haemophilus parainfluenzae* were detected. Significant differences in maternal faecal and human milk bacterial composition were maintained 4 to 8 weeks after the intervention. This pilot study demonstrates that short-term changes in maternal diet during lactation can alter the maternal faecal and human milk microbiota.

Citation: Sindi, A.S.; Stinson, L.F.; Gridneva, Z.; Leghi, G.E.; Netting, M.J.; Wlodek, M.E.; Muhlhausler, B.S.; Rea, A.; Trevenen, M.L.; Geddes, D.T.; et al. Maternal Dietary Intervention during Lactation Impacts the Maternal Faecal and Human Milk Microbiota. *Proceedings* **2023**, *93*, 22. https://doi.org/10.3390/proceedings2023093022

Academic Editors: Debra J. Palmer and Nicolas L. Taylor

Published: 16 January 2024

Keywords: diet; human milk microbiome; maternal faecal microbiome

Author Contributions: Conceptualization, L.F.S., D.T.G., M.E.W., B.S.M., M.J.N. and M.S.P.; methodology, A.S.S., L.F.S., G.E.L. and Z.G.; software, G.E.L. and A.S.S.; validation, A.S.S.; formal analysis, M.L.T. and A.R.; investigation, A.S.S.; resources, M.S.P.; data curation, M.L.T., A.R. and A.S.S.; writing—original draft preparation, A.S.S.; writing—review and editing, L.F.S., M.E.W., B.S.M., M.J.N., G.E.L., Z.G., M.L.T., A.R., D.T.G. and M.S.P.; visualisation, A.S.S.; supervision, L.F.S., D.T.G., M.E.W., B.S.M., M.J.N. and M.S.P.; project administration, L.F.S., D.T.G. and M.S.P.; funding acquisition, D.T.G. and M.S.P. All authors have read and agreed to the published version of the manuscript.

Funding: L.F.S., Z.G., and D.T.G. are supported by an unrestricted research grant from Medela AG (Switzerland), administered by The University of Western Australia. Umm Al-Qura University, Saudi Arabia provides a Ph.D. scholarship for A.S.S. M.S.P. is supported by a National Health and Medical Research Council Ideas grant (APP #2010530). The funding bodies had no role in the design of the study, collection/analysis/interpretation of data, writing of the manuscript, or in the decision to publish the results.

Institutional Review Board Statement: The study was conducted in accordance with the Declaration of Helsinki, and approved by The University of Western Australia Human Research Ethics Committee (RA/4/20/4953), and registered on the Australian New Zealand Clinical Trials Registry (ACTRN12619000606189).

Informed Consent Statement: Informed consent was obtained from all subjects involved in the study.

Data Availability Statement: Data described in the manuscript are freely available without restriction at https://www.ncbi.nlm.nih.gov/bioproject/PRJNA898341 accessed on 4 November 2022. All raw sequencing data have been deposited on NCBI's Sequence Read Archive (accession number: PRJNA898341).

Acknowledgments: We would like to acknowledge Kirsty O'Hehir for assistance with study recruitment and the implementation of the dietary intervention. We also would like to thank Ching Tat Lai for his help with statistical analysis.

Conflicts of Interest: D.T.G. declares participation in the Scientific Advisory Board of Medela AG. L.F.S., Z.G., and D.T.G. are supported by an unrestricted research grant from Medela AG, administered by The University of Western Australia. The funders had no role in the design of the study; in the collection, analysis, or interpretation of data; in the writing of the manuscript; or in the decision to publish the results.

proceedings

Abstract

Screening for Lactation Risk Factors in Pregnancy †

Stuart A. Prosser [1,2], Philip Vlaskovsky [3], Donna T. Geddes [2,4,5] and Sharon L. Perrella [2,4,5,*]

1 Western Obstetrics, Balcatta 6021, WA, Australia; stuart@westernobs.com.au
2 School of Molecular Sciences, The University of Western Australia, Crawley 6009, WA, Australia; donna.geddes@uwa.edu.au
3 Department of Mathematics and Statistics, School of Physics, Mathematics and Computing, The University of Western Australia, Crawley 6009, WA, Australia; philip.vlaskovsky@uwa.edu.au
4 ABREAST Network, Perth 6000, WA, Australia
5 UWA Centre for Human Lactation Research and Translation, Crawley 6009, WA, Australia
* Correspondence: sharon.perrella@uwa.edu.au; Tel.: +61-6488-1208

† Presented at the Australian Breastfeeding + Lactation Research and Science Translation Conference (ABREAST Conference 2023), Perth, Australia, 10 November 2023.

Keywords: breastfeeding; lactation; risk assessment; body mass index; gestational diabetes mellitus

The early cessation of exclusive or any level of breastfeeding is often attributed to low milk supply. While perceptions of low milk supply are not always correct, true low milk supply may result from inadequate and infrequent milk removal, as well as postpartum complications such as major postpartum haemorrhage. There is some evidence of associations between certain anatomical and endocrine factors with shorter durations of exclusive or any level of breastfeeding. Many of these factors are identifiable during pregnancy, yet traditional models of maternity care typically do not consider lactation risk. This study aimed to determine the prevalence of anatomical and endocrine lactation risk factors and early breastfeeding outcomes in a cohort of Australian women.

De-identified antenatal lactation risk screening and infant feeding data were extracted from clinical records at maternity clinics where lactation risk screening is routinely performed. The following risk factors were recorded: pre-pregnancy body mass index (BMI, with underweight, <18.5; with normal weight, 18.5–24.9; with overweight, 25.0–29.9; with obesity, ≥30.0), breast growth in pregnancy, pre-existing diabetes, gestational diabetes mellitus (GDM), breast hypoplasia, breast/nipple surgery, nipple piercing, polycystic ovary syndrome (PCOS), thyroid disorder and pituitary disorder. "No lactation risk" was defined as having a normal pre-pregnancy BMI and none of the listed risk factors. Feeding status at 6–8 weeks postpartum was classified as 'fully breastfeeding': no food or fluids other than breast milk, 'mixed feeding' both breast and commercial infant formula milk or 'formula only'. The relative risk for fully breastfeeding at 6–8 weeks (in relation to women with no antenatally identified risk factors) was calculated for each risk factor.

There were 581 records accessed, with complete breastfeeding screening data available for n = 408 and n = 160 missing one item (typically BMI or breast growth data). Two-hundred forty-three women (48.5%) had no identified lactation risk factors, 97 (19%) had one risk factor, and 161 (32%) had two or more risk factors. The most prevalent risk factors were pre-pregnancy BMI ≥ 25.0 (220 (42%) with overweight or with obesity), no breast growth 111 (24%), and GDM 66 (11%). Feeding outcome data were available for n = 414. Of women with no antenatally detected lactation risk factors, 81 (77%) were fully breastfeeding at 6–8 weeks postpartum. Relative risks of not fully breastfeeding at 6–8 weeks were obesity BMI (RR = 2.01 (1.44–2.81, p < 0.001)); overweight or obesity BMI (RR = 1.74 (1.30–2.34, p < 0.001)), the presence of two or more risk factors (RR = 1.76 (1.33–2.32, p < 0.001) and BMI ≥ 25.0 together with GDM (RR = 2.18 (1.58–3.01, p < 0.001).

Citation: Prosser, S.A.; Vlaskovsky, P.; Geddes, D.T.; Perrella, S.L. Screening for Lactation Risk Factors in Pregnancy. *Proceedings* **2023**, *93*, 24. https://doi.org/10.3390/proceedings2023093024

Academic Editors: Debra J. Palmer and Nicolas L. Taylor

Published: 21 March 2024

Lactation risk factors may be identified in half of all pregnant women, so lactation risk screening may offer an important strategy in providing targeted breastfeeding education and support to optimise breastfeeding outcomes.

Author Contributions: Conceptualization, S.A.P. and S.L.P.; methodology, S.L.P.; formal analysis, P.V. and S.L.P.; investigation, S.L.P.; resources, S.A.P. and D.T.G.; data curation, S.L.P.; writing—original draft preparation, S.L.P.; writing—review and editing, P.V., D.T.G. and S.A.P.; supervision, S.L.P.; project administration, S.L.P.; funding acquisition, D.T.G. All authors have read and agreed to the published version of the manuscript.

Funding: The salaries of D.T.G. and S.L.P. were funded by an unrestricted research grant from Medela AG (Switzerland), administered by The University of Western Australia. The funders had no role in the design of the study; in the collection, analyses, or interpretation of data; in the writing of the manuscript; or in the decision to publish the results.

Institutional Review Board Statement: The study was conducted in accordance with the Declaration of Helsinki and approved by the Human Research Ethics Committee of The University of Western Australia (2023/ET000333).

Informed Consent Statement: A waiver of consent was obtained from the Human Research Ethics Committee of The University of Western Australia (2023/ET000333).

Data Availability Statement: The data that support the findings of this study are available from the corresponding author upon reasonable request.

Conflicts of Interest: D.T.G. declares participation in the Scientific Advisory Board of Medela AG. D.T.G. and S.L.P. are supported by an unrestricted research grant from Medela AG, administered by The University of Western Australia. The funders had no role in the design of the study; in the collection, analyses, or interpretation of data; in the writing of the manuscript; or in the decision to publish the results. All other authors declare no conflicts of interest.

MDPI
St. Alban-Anlage 66
4052 Basel
Switzerland
www.mdpi.com

Proceedings Editorial Office
E-mail: proceedings@mdpi.com
www.mdpi.com/journal/proceedings

www.ingramcontent.com/pod-product-compliance
Lightning Source LLC
Chambersburg PA
CBHW041955260926
51850CB00019B/2926
* 9 7 8 3 7 2 5 8 0 7 8 7 1 *